Sustainability and Innovation

Jens Horbach (Editor)

Indicator Systems for Sustainable Innovation

With 22 Figures and 9 Tables

Physica-Verlag
A Springer Company

Professor Dr. Jens Horbach
University of Applied Sciences Anhalt
Strenzfelder Allee 28
06406 Bernburg
horbach@wi.hs-anhalt.de

Cataloging-in-Publication Data
Library of Congress Control Number: 2004116573

ISBN 3-7908-1553-5 Physica-Verlag Heidelberg New York

Physica is a part of Springer Science+Business Media

springeronline.com

Printed in Germany

Softcover-Design: Erich Kirchner, Heidelberg

SPIN 11014492 88/3130-5 4 3 2 1 0 – Printed on acid-free paper

Preface

During the last thirty years the emissions of many pollutants such as sulphur dioxide have been reduced considerably. Besides structural change between branches environmental innovations can be identified as the main driving force of this positive development. Nevertheless, there are many remaining and even new pollution sources so that a sustainable development requires further technical, social and institutional innovations.

The research programme “Framework conditions for innovation towards sustainability :[riw]” initiated by the German Federal Ministry of Education and Research tries to provide the knowledge on the significance and effects of framework conditions on the generation of innovations that are able to contribute to the realisation of a sustainable development. Within this programme, including more than twelve different research projects, an international working group has been established. The primary target of this working group was to develop an indicator system allowing to evaluate sustainable effects of (environmental) innovations by linking the different experiences of the participating projects with the international theoretical and empirical research on indicators for sustainable innovation.

This volume contains the main research results of the working group. Besides more conceptually oriented articles the focus lies on the development of indicators for specific environmental innovation systems.

The editor gratefully acknowledges the financial and organisational support from the German Federal Ministry of Education and Research and the Project Management Office for Environment and Climate at the GSF-National Research Centre for Environment and Health in Munich. Furthermore I am grateful to Eberhard Feess, Jens Hemmelskamp, Joseph Huber and Marco Lehmann-Waffenschmidt for the admission of this volume in the Springer series "Sustainability and Innovation - Nachhaltigkeit und Innovation" and to Ilja Karabanow and Selina Pohl for their help in proof-reading and technical edition.

Jens Horbach

Contents

Preface V

Methodological Aspects of an Indicator System for Sustainable Innovation 1
Jens Horbach

The Innovation Effects of Environmental Policy Instruments 21
Nick Johnstone

Indicators for a Sustainable Technology Development - A Dynamic Perspective 43
Christian Sartorius

Indicators for Lead Markets of Environmental Innovations 71
Marian Beise and Klaus Rennings

Policy Impacts on Macroeconomic Sustainability Indicators when Technical Change Is Endogenous 95
Joachim Schleich, Rainer Walz, Bernd Meyer and Christian Lutz

Indicators for Innovations Towards Prevention of Toxic Risks 107
Andreas Ahrens and Arnim von Gleich

Environmental Innovations in the Chemical Industry - Case Studies in a Historical Perspective 125
Melanie Monßen

A Sustainable Innovation Scorecard for the Electronics Industry Innovation System 143
Joachim Hafkesbrink and Marianne Halstrick-Schwenk

Water Management Towards Sustainability - An Indicator System to Assess Innovations 179
Hartmut Clausen and Joachim Hafkesbrink

List of Contributors 205

Methodological Aspects of an Indicator System for Sustainable Innovation

Jens Horbach

1 Introduction

The aim of this paper consists in the development of a framework for an indicator system for sustainable innovation. This framework serves as a basis for the following articles in this book containing detailed indicator systems for various environmental innovation fields that have been regarded within the German :[riw] programme (research initiative on sustainability and innovation). The :[riw] initiative of the German Federal Ministry of Education and Research aims to "… provide the knowledge on the significance and effects of framework conditions on the generation of innovations that could contribute to the realisation of a sustainable development." (Grablowitz and Hemmelskamp 2001).

Firstly, the development of an indicator system requires a definition of the term "sustainable innovation". The main task of this paper consists in discussing the relevant dimensions of sustainable innovation from a theoretical perspective. Furthermore, the problems of the availability of data for indicator systems have to be discussed. Section 6 contains a synopsis of the different examples for indicator systems in the involved :[riw] projects.

2 Definitions of sustainability and sustainable innovation

The term sustainable innovation has not yet been sufficiently defined in the literature. This is due to the fact that both components of this term are objects of a very broad discussion. In the following this debate will be summarised separately for sustainability and innovation. In a second step these definitions will be combined.

Definitions of sustainability

Following the Brundtland report sustainability describes "... a process of change in which the exploitation of resources, the direction of investments, the orientation of technological development and institutional changes are made consistent with future as well as present needs." (WCED 1987). In other words sustainability means a "development that meets the needs of the present generation without compromising the ability of future generations to meet their own needs" (Callens and Tyteca 1999). An operational concept to characterise sustainability has been developed by Daly. His management rules can be summarised as follows (Constanza et al. 2001 or Hanley et al. 1997):

- Renewable resources: the harvest rate is not allowed to exceed the regeneration rate of the resource;
- Pollution: pollution and waste production is not allowed to fall below the assimilation capacity;
- Non-renewable resources: the depletion of the resource has to be accompanied by the creation of a substitute.

Especially the last one can pose problems because the optimism of the so-called Hartwick rule which says that "The stock of capital could be held constant by reinvesting all Hotelling rents from non-renewable resource extraction in man-made capital" (Hanley et al. 1997) (concept of "weak sustainability") is not realistic in all cases. Beyond other arguments, the Hartwick rule has been criticised because the substitutability of natural resources by man-made capital is not always possible. The representatives of the concept of strong sustainability even assume that natural and man-made capital are complements. Furthermore, when individuals derive utility directly from the environmental capital stock non-declining consumption is not automatically equivalent to non-declining welfare over time (Hanley et al. 1997).

A practicable concept of sustainability has to combine elements of the two concepts (see also Rat von Sachverständigen für Umweltfragen 2002). We can depart from the assumption of the so-called weak sustainability supposing substitutability between the natural and the "artificial" capital stock. It is not possible to stop oil exploitation abruptly and to use other energy sources but it has to be the task of environmental (or sustainable) innovation to make a substitution possible. But in many cases it will be very uncertain whether adequate substitutes can be found. "Multifunctionality in conjunction with uncertainty and with some new findings in the limited elasticity of substitutability as well as the aesthetic qualities and social ("biophilic") amenities of unspoiled nature provide some sound patterns of arguments against weak sustainability."(Ott and Döring 2002) For that reason it is useful to follow the precautionary principle in environmental politics and to use for instance "safe minimum standards" with respect to natural capital to avoid restraints resulting from a low elasticity of substitution between natural and artificial capital (see also Hampicke 1992).

Invention and innovation

The term invention mainly describes the idea and first development of a new process or product whereas innovation deals with extending this idea to further development and utilisation of a product or process. Innovations will only be successful if they provide us with a certain service which we could not use before or if they improve an existing service and/or if they lead to lower cost. In the literature, five stages of the innovation process are distinguished: Recognition - Invention - Development - Implementation - Diffusion (see e. g. Grupp 1997).
Furthermore, innovations can be classified into product- and process innovations, organisational and institutional innovations (see also sections 4 and 5).

Environmental and sustainable innovation

Environmental and sustainable innovations represent subsets of innovation systems. However, they can not be separated from these systems in all cases. Following a definition of Kemp, Arundel and Smith (2001) "Environmental innovation consists of new or modified processes, techniques, systems and products to avoid or reduce environmental damage". Sustainable innovations can only be successful if they allow for the same use value at lower environmental cost. For successful market diffusion, it may even be necessary that the new product has an even higher use value (in addition to lower environmental damage) (Kemp et al. 2001).

Sustainable innovations not only comprise the environmental dimension but also economic, social and institutional aspects. They improve the realisation of the aims of a sustainable development and represent a subset of all innovations. Because of the complexity of the ecological, economic and social system a simultaneous consideration of all levels of sustainability will not be possible so that we have to concentrate on sustainable innovations aiming to reduce environmental impacts. Nevertheless, we have to consider the economic, social and institutional consequences of sustainable innovations when constructing an indicator system.

3 Relevant dimensions of sustainable environmental innovation from a theoretical perspective

Even if we focus on ecological sustainability, an analysis of sustainable innovations requires a broad comprehension of the whole innovation system including determinants, description of the innovation itself and ecological, economic and social impacts (see also table 1). Besides the knowledge about the innovation itself an indicator system has to consider the underlying forces of the innovation so that we can learn how to create incentives for more innovation activities by adequate (environmental) policy measures. Furthermore, the sustainability of an innovation can only be evaluated if we know something about its impacts.

The aim of this section consists in discussing the theoretical foundations and problems of these different dimensions of sustainable innovation.

Table 1. Levels of analysis of an indicator system

Level of analysis	Examples
Determinants of sustainable innovation	Market demand, „fitting“ time window, environmental policy measures, path dependencies
Description of the innovation	Product and process innovations, organisational and institutional changes, end-of-pipe versus integrated environmental innovations
Ecological, economic and social impacts	Emission reductions, income distribution, employment effects

Determinants (and obstacles) of sustainable innovations

For a long period of time, neo-classical economists stated that technical progress is exogenous and can therefore not be explained. Contrary to the simple neo-classical model, new institutional, evolutionary economics and the new growth theory made attempts for an endogenisation of technical progress. Up to now, these theoretical developments did not lead to a comprehensive theory for explaining (sustainable) innovation. Nevertheless, it is possible to derive the main driving forces of sustainable innovation from different theoretical explanations. In a second step these determinants have to be translated in quantifiable indicators.

The main incentive for the innovation activities of a firm consists in getting higher market shares and more profit with respect to competitors. Therefore it is very important to analyse the specific market conditions of sustainable innovation because firms are only interested in spending money for innovation activities if they succeed in capturing the innovation rents of these activities.

Hence we have to include variables like the contestability of the corresponding market. E. g. the "... existence of vested interests among individuals (firm owners, managers, workers, researchers and so on) specialised in the old technologies who therefore are tempted to collude and exert political pressure in order to delay or even prevent the arrival of new innovations that might destroy their rents" (Aghion and Howitt 1998) has to be considered.

Table 2. Determinants of sustainable innovation

Supply side	• offsets (cost savings) caused by environmental innovations • market characteristics: company size and market structure (monopolistic structures can prevent environmental innovations because there is no incentive to innovate at all) • possibilities for the protection of innovations (problem of internalising the positive externalities (spillovers) of an innovation); attitude to risk and the uncertainty of environmental innovations: „Asymmetric information and moral hazard are notable characteristics of the initial situation of potential users of an innovation." (Klemmer et al. 1999) • path dependencies: the available technological possibilities (accumulation of human capital, available knowledge) induce further innovations (technology push hypothesis, see also Hemmelskamp 1999) • fitting time window for the realisation of the innovation
Demand side	• market demand (demand pull hypothesis): state, consumers and firms • social awareness of the need for clean production; environmental consciousness and preference for environmental friendly products
Institutional and political influences	• environmental policy (incentive based instruments or regulatory approaches) • institutional structure: e. g. political opportunities of environmentally oriented groups, organisation of information flow, existence of innovation networks • pressures from "the world community": e. g. international agreements on CO2

With respect to sustainable and environmental innovation, innovation policy plays a very important role because of the existence of externalities inducing market failures. This is particularly relevant for the initial phase of an innovation. In later phases, when a specialised human capital and adequate institutions have been established, the early developed innovations will promote further sustainable innovations. These path dependencies initiated by policy measures can even lead to economic and social advantages when other countries also begin to demand sustainable technologies. This is one of the main components of the so-called Porter hypothesis (Porter and van der Linde 1995). The new growth theory enforces this argument stating that in regions characterised by a high capacity of innovative firms positive external effects can attract other innovating firms.

Specifically, environmental innovations are often characterised by high risks and uncertainty. Therefore the organisation of financing R&D in this field is of interest. Concerning an indicator system we have to ask e. g. whether there are possibilities to obtain "venture capital". Sustainable innovations can be enforced by a fitting design of subsidies and patent legislation by the government.

A very important element of the success of sustainable innovations coming from evolutionary economics consists in the "windows of opportunity" idea (see also the contribution of Sartorius in this volume). An innovation can only be successful if other factors and circumstances are fulfilled at the same time. If a process innovation appears when the relevant branch has just invested in another technology it will probably not be used because of sunk costs. Furthermore, the preferences of consumers decide on the diffusion of new innovative environmentally friendly products. These examples show that an analysis of success factors of sustainable innovations that have to be integrated in an indicator system requires a profound description of time dependent factors. From a policy perspective it is favourable to create an innovation friendly environment.

Description of environmental innovation

The main problem of the description of the environmental innovation itself consists in identifying the "subset environmental innovation" (Klemmer et al. 1999). In the case of innovations connected with end-of-pipe measures this task will be easy but if we regard the so-called integrated environmental innovations, the environmental or sustainable "part" of the innovation is not separable from the whole innovation system (see also the section dealing with data problems). This requires a totally different set of indicators measuring the sustainability of innovations because we have to include all actors contributing to the innovation. This can be very complicated because integrated environmental technologies are often developed in a close cooperation between manufacturers, users or suppliers of complementary products or inputs (Halstrick-Schwenk et al. 1994).

Another dimension which has to be considered in an indicator system is the description of the different phases of innovation (Klemmer et al. 1999) departing from the generation of new knowledge (invention), the spread or first application (adaptation) and the diffusion as new product, process, institutional arrangement or behaviour.

Ecological, economic and social impacts

Sustainability implies that we have to search for a dynamic Pareto-optimum accounting for the ecological, economic and social components of the utility functions of all members of present and future generations. Because of the complexity of this task this is only a theoretical demand but an indicator system has to have at least the aim to fulfil these conditions. Another problem arises because the different impacts cannot be regarded separately. For instance, it is not meaningful to ameliorate the environmental quality if the avoidance costs exceed the respective utility. For that reason adequate indicators have to be able to describe the interactions between the different dimensions of sustainability.

As a starting point for a description of the relationship between ecological and economic impacts we can use the concept of the Environmental Kuznets Curves (EKC) as an overall indicator for the sustainability of innovations when we succeed in separating (econometrically) the impact of innovations besides other determinants of EKC´s. The EKC postulates an inverted u-shaped curve representing the relationship between important pollutants and per capita GDP analogous to the relationship between income inequality and income per capita that has been analysed by Kuznets (1955). The EKC describes a time path characterised initially by a positive correlation between pollutants and GDP becoming negative during a later phase of the economic development of a country. If the pollution level from which we start is already sustainable the downsloping part of the EKC indicates a sustainable development with respect to the concerned pollutant. In the literature it could be proven that innovation activities are the main sources of the appearance of EKC´s besides inter-sectoral structural change (de Bruyn 2000).

Apart from the impacts of environmental innovation on GDP other important examples of relevant economic and social variables are changes in productivity, employment or income distribution. At last, the final choice of variables to be used in an indicator system depends on the specific problem or pollutant.

A great problem assessing the different impacts of sustainable innovations results from the fact that the evaluation and hence the comparison of different indicators is very difficult. What would be better? A ton of avoided SO_2 emissions or 200 people unemployed? Such evaluation problems can be recognised

- during the selection and the weighting of the indicators;
- with respect to the interaction between the indicators;
- when interpreting the results of the developments described by the indicators.

Finally, it is not possible to solve these evaluation problems but in most cases it is appropriate to use indicators describing changes of parameter values instead of levels.

4 Concepts of indicator systems

This section contains a brief overview of the most important indicator systems of sustainability (see e. g. Sors 2001; Endres and Radke 1998; Constanza et al. 2001; OECD 1998). Furthermore, we have to discuss which indicator system can easily be linked and widened with respect to environmental innovation systems.

In the literature, one- and multi-dimensional indicators are distinguished:

One-dimensional indicators:

- Eco-social product: This concept tries to correct the social product by taking environmental expenditures into consideration;
- Non-declining per capita human well-being over time (Constanza et al. 2001): The indicator requires that the overall capital stock is not allowed to decline. To fulfil this condition weak sustainability has to be assumed postulating substitutability between natural and produced capital;
- Index of Sustainable Economic Welfare (ISEW): Amelioration of the concept of the eco-social product by considering the sustainability of consumption and the effects of consumption on natural capital and income distribution (Constanza et al. 2001).

Multi-dimensional indicators:

- The Pressure-State-Response approach (PSR) includes the pollution caused by consumption and production activities, the state of the environmental indicators and the reactions of the society with respect to changes of the environmental quality. This approach neglects the driving forces of sustainability;

- The Driving Force-State-Response (DSD) approach of the Commission on Sustainable Development (CSD) of the United Nations enlarges the PSR approach by taking the driving forces of pressures on the environment into account: This approach shows a matrix structure of important indicators for the driving forces, the state and the societal responses of sustainable development by regarding ecological, economic, social and institutional dimensions (Constanza et al. 2001). The problem is that the links between the different variables are not analysed. This has to be done if we want to investigate the impacts of sustainable innovations.

If we want to connect indicators of sustainable innovation with general indicator systems of sustainability it would be favourable to use the DSD system because this approach includes the driving forces and the relevant dimensions (ecological, economic, social and institutional) that have to be considered for sustainable innovation.

5 Data problems of indicator systems

In the following, the availability of adequate data for the description of environmental innovation systems is discussed. This is a very important issue because there are many difficulties in getting fitting data. We have to avoid the problem that the available data itself determines the indicator system.

The focus is on the description of the innovation system. Concerning the heterogeneous determinants and impacts we have to look for specific data sources.

In general the following types of data sources for the description of environmental innovations are available (see e. g. Kemp et al. 2001; Grupp 1997):

- Official statistics: E. g. PACE (Pollution Abatement Costs and Expenditures);
- Surveys and case studies;
- Literature-based environmental innovation surveys (LBIO) using entry forms for environmental awards, surveys of successful environmental innovations, articles in the most important trade journals. A problem of these indicators is that the results may be biased towards the most successful environmental innovations. Furthermore, only innovations which are identifiable as environmental innovations can be taken into consideration;
- Patent statistics;

- Indicators derived from the monitoring of environmental management systems (Examples: EMAS: Eco Management Audit Systems or ISO 14001 scheme for the certification of corporate environmental management at the firm level, CER: Corporate Environmental Reporting). The indicators resulting from the management systems are e. g.: visible evidence of abatement efforts, reported emission rates, self-assessed environmental performance, involvement in a voluntary environmental programme (Johnstone 2001).

Table 3 shows a classification of environmental innovations with examples of indicators and the main data sources. All categories of environmental innovations can be analysed by surveys and/or case studies but this way of proceeding is very costly and therefore not always adequate for the monitoring of sustainable innovations. Especially for a long-term observation of developments we need indicators which can be permanently derived from existing sources like official statistics, patent statistics or bibliometric data.

With respect to process integrated measures severe data problems may occur because it is difficult to separate the environmental part from the whole innovation system. In this case, we need mainly to refer to surveys and case studies or we may use output indicators like the development of energy consumption with respect to a particular production process.

Table 3. Data sources for measuring environmental innovation activities

Types of environmental innovation	Main indicators	Main data sources
Eco Products (goods and services)	Output indicators: Environmental characteristics of the new products with respect to comparable products (by using eco balances), patents Input indicators: R&D expenditure and personnel	Surveys and information from industrial associations, eco labels, case studies, patent statistics, bibliometrics
Process integrated measures, logistics, organisational measures	Improvement of energy intensity, reduction of material use etc.	Surveys, case studies, official statistics
End-of-pipe processes	Innovation activities of suppliers of environmental goods (patents, R&D expenditure and personnel) Demand side: analysis of environmental investment	Surveys, official product statistics, official statistics of environmental investment, patent statistics, case studies, bibliometrics
Recycling	Recycling share with respect to total waste amount	Official statistics (recycling belongs to the ISIC or NACE classification), surveys, patent data, bibliometrics, case studies

Modified classification from Rennings and Zwick (2001), own considerations.

6 Contributions of different :[riw] projects to an indicator system

The following overview of the different innovation systems included in this volume demonstrates that it is not useful to construct a single indicator system for different sustainable innovations. The heterogeneity of environmental innovations (see also the preceding sections) only requires a common structure of an indicator system as developed in this paper. The specific indicators have to be closely linked to the corresponding innovation system. The contributions of this volume provide examples of such indicators for different innovation systems and with different emphases on determinants, description of the innovation system and impacts.

In the following, a synopsis of the different innovation systems with respect to an indicator system will be presented.

Johnstone develops general requirements for indicator systems of sustainable innovation. He emphasises the aspect that not only the rate of innovation matters but also the direction of innovations. Following his definition an innovation direction is optimal if it "is cost-minimising in the long-run with respect to the realisation of the given environmental objective".

The main interest of the SUSTIME (Sartorius) project consists in analysing the determinants of a successful innovation process in general and with regard to sustainable technologies in particular. The author discusses theoretical aspects of the relevance of time for the realisation of innovations. These aspects consider the diffusion of any sustainable technology as the result of the complex interaction between a society's techno-economic, political and social subsystems.

The diffusion phase of new environmental friendly (and sustainable) products is analysed in the LEADMARKET (Rennings, Beise) project. As examples, the authors regard fuel-efficient passenger cars or the diffusion of wind energy. Detailed indicators for the determinants of these innovations are developed.

The project INNOMOD (Schleich, Walz, Lutz, Meyer) develops indicators for process innovations and process-integrated measures in the steel industry. The authors provide quantitative indicators for the whole system determinants - innovations – impacts that can be (and are used by the authors) for econometric analysis.

The SUBCHEM (Ahrens, von Gleich) project develops indicators for substitution processes related to hazardous substances in products and processes in the chemical industry. The empirical data is based on 12 case studies.

The COIN (Monßen) project develops indicators for the determinants of product, end-of-pipe and organisational innovations in the chemical industry. The indicators are derived from an analysis of historical firm data.

The INVERSI (Hafkesbrink, Halstrick-Schwenk) project has a special focus on institutional and organisational innovations. An extensive indicator system has been developed including determinants, the innovation system and impacts with respect to innovations in the field of waste disposal in connection with globalisation and electronic markets.

The AQUASUS (Clausen, Hafkesbrink) project derives indicators for the sustainability of water use. The innovation system concentrates on end-of-pipe and organisational innovations. The analysis contains a very broad analysis of the determinants of the considered innovations.

Table 4. Contributions of the different riw-projects

a) Time strategies of an ecological innovation policy (SUSTIME) (Sartorius)	
Phase and character of sustainable innovations	
Focus of this contribution lies on the analysis of the determinants of sustainable innovations becoming effective. Application of the indicator system is illustrated for the phase-out of CFC and the development of corresponding substitutes.	
Determinants of in-/stability in the ...	*Examples for indicators*
Techno-economic system	Economies of scale/scope, network and learning effects, degree of competition, existence of niche markets, identification of investment cycles
Political system	Subsidies, norms, standards (often in disfavour of more radical innovations), stability of majorities in parliament, corporate structure, influence of interest groups, re-submission cycles, relevance of supra-national structures etc.
Socio-cultural system	Initial recognition of the environmental problems in the scientific literature, formation of social attitudes with regard to environmental problems; media attention as amplifier of social attitude

Table 4. (cont.)

b) Lead markets of environmental innovations (LEADMARKET) (Beise, Rennings)	
Phase and character of sustainable innovations Diffusion phase of product innovations	
Determinants	*Examples for indicators*
Env. regulations, policy diffusion	Eco-taxes
Demand trends	Demand of households for env. friendly products
Demographic trend	Age structure of the population
Pressing environmental problems	Development of CO_2 emissions
Prices	Factor cost trends
Export advantage	RCA values
Market structure advantage	Market size
Innovation system	*Examples for indicators*
Fuel-efficient Passenger Cars Wind energy	Actor configuration, networks
Impacts	*Examples for indicators*
Economic success, diffusion	Share of diesel motors with direct injection International diffusion, world market shares of wind manufacturers

c) Innovations and emissions of air pollutants (INNOMOD) (Schleich, Walz, Lutz, Meyer)	
Phase and character of sustainable innovations Focus on development and diffusion, process innovations, process-integrated measures	
Determinants	*Examples for indicators*
Determinants of technical progress (prices, input-structure, market-structure, policy changes, cost-push hypothesis)	Expected price variables R&D expenditure Herfindahl-Index, share of imports Scenarios of policy changes
Determinants of diffusion through investments	Output and existing production capacities Relative input prices for case study sector Real interest rate
Innovation system	*Examples for indicators*
Steel production	Changing input structures, energy input
Impacts	*Examples for indicators*
Economic	GDP, employment, investments, government budget
Ecological	CO_2-emissions, energy consumption
Social	Income distribution, qualitative employment effects

Table 4. (cont.)

d) Sustainable substitution of dangerous chemicals (SUBCHEM) (Ahrens, von Gleich)	
Phase and character of sustainable innovations	
Product and process innovations, organisational and institutional innovations	
Determinants	*Examples for indicators*
Forms of cooperation	Existence of tools (agreed among suppliers and users) for comparative evaluation of products
Institutional structure of information flows	Extent and quality of risk related information available in the market by legal requirement and rate of compliance Gaps in responsibility during life cycle by legal requirement and corporate commitment
Innovation system	*Examples for indicators*
Product innovation	Rate of new substances placed on the market and sectors of use Turnover in active substances in the pool of existing substances in the market Profit per kg of chemical
Organisational innovation	Existence of a corporate policy and reporting systems related to toxic risks at company level Purchase of chemicals includes comparative risk assessment as a management routine
Impacts	*Examples for indicators*
Economic	Amount of substance need to provide a certain unit of benefit
Ecological	Rate of certain occupational diseases Breeding success of birds Occurrence of persistent substances in the environment

Table 4. (cont.)

e) Cooperative institutions in the chemical industry (COIN) (Monßen)	
Phase and character of sustainable innovations	
Product innovations, end-of-pipe and organisational and institutional innovations	
Determinants	*Examples for indicators*
Environmental regulation	Regulation of waste water disposal
Influence of interest groups	Complaints of fishermen
Innovation system	*Examples for indicators*
Product innovation	Polyaspartic acid
End-of-pipe	Dilute acid
Organisational innovation	Water commission
Impacts	*Examples for indicators*
Economic	Cost reductions
Ecological	Reduction of pollutant concentrations (sulphate, pigments) Changes in the water quality of the North Sea Changes in fish population Changes in sulphuric acid concentration Alternative methods of waste disposal Increase of resource efficiency

Table 4. (cont.)

f) Globalization and electronical markets in the field of waste disposal (INVERSI) (Hafkesbrink, Halstrick-Schwenk)	
Phase and character of sustainable innovations	
End-of-pipe and organisational and institutional innovations	
Determinants	*Examples for indicators*
Environmental (or other) regulations	Prohibition of the use of substances
Consumer demand	Purchasing of green TV´s
Existing technologies	New recycling techniques
Competition situation and market structure	Firm structure, concentration
Innovation system	*Examples for indicators*
Process innovations	Use of resources
Product innovations	Environmental quality of products (e.g. energy consumption when using the product)
Organisational changes within firms	Introduction of management systems (eco-audits)
Organisational changes between firms	Taking back systems
Impacts	*Examples for indicators*
Economic	Cost efficiency of taking back systems Development of costs of waste disposal Development of competitiveness
Ecological	Development of material and energy use
Social	Emissions, recycling shares

Table 4. (cont.)

g) Sustainable water management (AQUASUS) (Clausen, Hafkesbrink)	
Phase and character of sustainable innovations	
End-of-pipe and organisational and institutional innovations	
Determinants	*Examples for indicators*
Institutional and legislative context: Environmental regulation on EU level, national level, requirements of authorities, requirements of stakeholders	Selectivity of regulation Dynamic incentive impact of driver Strength of inter-actor relations Selectivity/specificity of public funding
Market and Technology: Competition/Non-Competition, demand, technology push, cost-pressure	Availability of new (basic) technologies Volume of drain renovation needs
Other drivers: Nature area conditions (i.e. topography) Information/public opinion	Limit values or thresholds for emissions Topography Barometer of public opinion
Innovation system	*Examples for indicators*
Complexity of actors configuration	Linkages between actors (market and non-market linkages)
Behavioural attitudes of innovation players	Co-ordination of innovation actors Size and vertical integration of industry ownership Situation of innovation actors
Diversity of innovation sources	Importance of different innovation/information sources
Dissemination speed of innovation	Time window for adaptation and diffusion
Technical innovation altitude	Complexity of technological solutions
Risk and uncertainty of implementation of innovation	Number and range of innovation actors involved and persons affected by innovations
Institutional arrangements	Influencing control of professional associations on political decision making
Impacts	*Examples for indicators*
Economic	Drinking water tariffs, degree of cost coverage, micro-economic efficiency, macro-economic efficiency, systems flexibility
Ecological	Keeping of limit values in water quality, Water trade, water balances, water extraction charges
Social	Discharge of persistent substances
Sustainability dimension (time variable)	Consumption per capita, use in circulation

7 Summary

In this paper a framework has been developed how to construct an indicator system for sustainable innovation. These innovations not only comprise the environmental dimension but also economic, social and institutional aspects. They contribute to the realisation of the aims of a sustainable development and represent a subset of all innovations. An indicator system for sustainable innovations which is relevant for political measures not only requires the description of the innovation system but also the determinants and the ecological, economic and social impacts of the respective innovations. Problems result from difficulties to identify the interactions between different indicators and from the evaluation of indicator values.

Furthermore, data problems have been discussed. In many cases - especially for process-integrated innovations - a fitting data set can only result from surveys or case studies.

A synopsis of different riw-projects contributing to an indicator system has shown that the framework developed in this paper can be applied to all the involved projects but in detail it is not useful to construct a single indicator system for all sustainable innovations e. g. comparable to the CSD approach. The variety and the complexity of innovation systems require specific indicators for each system.

References

Aghion P, Howitt P (1998) Endogenous Growth Theory. Massachusetts

Callens I, Tyteca D (1999) Towards indicators of sustainable development for firms - A productive efficiency perspective. Ecological Economics 28: 41-53

Constanza R, Cumberland J, Daly H., Goodland R, Norgaard R (2001) Einführung in die Ökologische Ökonomik. Stuttgart

De Bruyn S (2000) Economic growth and the environment: an empirical analysis. Dordrecht

Endres A, Radke V (1998) Indikatoren einer nachhaltigen Entwicklung: Elemente ihrer wirtschaftstheoretischen Fundierung. Berlin

Grablowitz A, Hemmelskamp J (2001) Research for new framework conditions towards sustainability. In: Hemmelskamp J (ed) Research Initiative on Sustainability and Innovation. München

Grupp H (1997) Messung und Erklärung des technischen Wandels: Grundzüge einer empirischen Innovationsökonomik. Berlin

Halstrick-Schwenk M, Horbach J, Löbbe K, Walter J (1994) Die umwelttechnische Industrie in der Bundesrepublik Deutschland. Essen, Halle

Hampicke U (1992) Ökologische Ökonomie, Individuum und Natur in der Neoklassik, Natur in der ökonomischen Theorie: Teil 4. Opladen

Hanley N, Shogren JF, White B (1997) Environmental Economics – In Theory and Practice. New York

Hemmelskamp J (1999) Umweltpolitik und technischer Fortschritt. Heidelberg

Johnstone N (2001) The Firm, the Environment, and Public Policy, final report of the OECD's "Eco-Efficiency Project". Paris

Kemp R, Arundel A, Smith K (2001) Survey indicators for environmental innovation (paper presented to conference „Towards Environmental Innovation systems“ in Garmisch-Partenkirchen)

Klemmer P, Lehr U, Löbbe K (1999) Environmental Innovation - Incentives and Barriers. Berlin

Kuznets S (1955) Economic growth and income inequality. American Economic Review 49: 1-28

OECD (Organization of Economic Cooperation and Development) (1998) Towards Sustainable Development - Environmental Indicators. Paris

Ott K, Döring R (2002) Strong Sustainability and Environmental Policy in Germany - The position of the German Council of Environmental Advisers on strong sustainability and its implementation in practical policy. Zürich

Porter ME, van der Linde C (1995) Towards a New Conception of the Environment - Competitiveness Relationship. Journal of Economic Perspectives 9: 97-118

Rat von Sachverständigen für Umweltfragen (2002) Umweltgutachten 2002 - Für eine neue Vorreiterrolle. Bundestagsdrucksache 14/8792. Berlin

Rennings K, Zwick T (2001) The employment Impact of Cleaner Production on the Firm Level: Empirical Evidence from a Survey in Five European Countries. ZEW-discussion paper Nr. 01-08. Mannheim

Sors JC (2001) Measuring Progress Towards Sustainable Development in Venice: A Comparative Assessment of Methods and Approaches. Fondazione Eni Enrico Mattei (FEEM) Nota Di Lavoro 16. Milano

WCED (World Commission on Environment and Development) (1987) Our Common Future. London

The Innovation Effects of Environmental Policy Instruments

Nick Johnstone

1 Introduction

The importance of the environmental policy framework in bringing about a technological trajectory which is less environmentally damaging has been noted. Indeed, as far back as the mid-1970s it was pointed out that "over the long haul, perhaps the most important single criterion on which to judge environmental policies is the extent to which they spur new technology towards the efficient conservation of environmental quality." (Kneese and Schultz 1975).

In this report some of the theory and evidence about the innovation effects of standard environmental policy prescriptions - whether they be economic instruments (emission taxes, tradable permits), direct forms of regulation (performance standards, emission limits, technology-based standards) or non-mandatory measures (voluntary agreements, information schemes) - will be reviewed.

In particular it will be argued that rather too much attention has been focussed on the effects of different instruments on the rate of innovation and rather too little on the direction of innovation. In particular, the report will review the implications of a number of factors which complicate the assessment of the innovation effects of different instruments: missing markets for certain environmental attributes of innovation; technological market failures; the point of incidence of the environmental policy; and, the existence of joint production of emissions.

The issue of direction is particularly important as it is arguably much easier for a policymaker to increase the rate of innovation, than it is to ensure that it is directed in the socially optimal manner - i.e. in a manner which is cost-minimising with respect to the attainment of a particular environmental objective in the long run. For various reasons government efforts to encourage innovation may succeed in increasing the rate of change, but in a manner which is not optimal with respect to the direction. As recent work on "lock-in" and network externalities has shown, the long-run costs of "misdirection" of the direction of innovation can be considerable.

2 The "orthodox" view of the innovation effects of alternative environmental policy instruments

Different policy instruments will affect the incentives for firms and households to develop and adopt environmentally-beneficial technologies in different ways; taxes and permits will affect the relative price of different factor inputs, encouraging firms to save on those factors which are closely linked to environmental damages; performance standards will place binding quantitative limits on the use of particular inputs or generation of particular emissions; technology standards or input bans will directly constrain the choice of technologies which can be used; and, information-based measures will affect the firm or household's perceptions of the relative merits of alternative choices of production processes or product design.

Much of the theoretical literature in this area concerns a comparison of market-based instruments relative to direct forms of regulation. While the case for market-based instruments (taxes, permits, deposit-refund schemes, etc....) relative to direct regulation (technology-based controls, performance standards, input bans, etc.....) has usually been made in static terms, at the theoretical level it is thought that the case is even more convincing when the dynamic effects in terms of technological innovation are examined. In particular, it is argued that the rate of change is more likely to be optimal since a greater proportion of benefits of technological innovation and adoption are realised by the firm itself under market-based instruments than is the case for many direct forms of regulation. Moreover, since market-based instruments are not "prescriptive" they are more likely than many types of direct regulation to ensure that the direction of technological change is cost-minimising with respect to the avoidance of damages (see Downing and White 1986; Milliman and Prince 1989; Nentjes and Wiserman 1987; Jung et al. 1996).

This stark juxtaposition of the technological effects of market-based instruments and more direct forms of regulation is somewhat of a caricature. In the first instance, it is clear that there are important differences between types of direct regulation. For instance, while a technology-based standard will provide little incentive to innovate, a performance-based measure will provide strong incentives for innovation and diffusion of technologies which achieve given environmental standards at lower financial cost[1]. In effect, under a technology-based regime, the potential innovator faces both

1 Although incentives will still be less than under most market-based instruments, since savings will only arise up to the point at which the standard is met.

a commercial risk and a regulatory risk, while under a performance-based regime only the former risk is important. Thus, unless the innovator believes that through innovation it can bring about a change in standards, firms will be less likely to innovate under technology-based systems[2].

However, the difference between the two can be overstated since in practice many performance-based standards are equivalent to technology-based standards, with the regulator only granting permits to particular technologies. Moreover, even if the regulator permits without delay all technologies which meet the performance standard, those technologies which are potentially more efficient (environmentally and financially) in the medium-run or long-run will still have zero share of the market until they meet the prescribed standard. (see Environmental Law Institute (1998) for some American examples). This vastly increases start-up costs and prevents supplier-user interactions which usually cut down development costs overall.

An input ban (or the prospect of such a ban) can provide a very important spur to the development of substitute materials. This case is often made with reference to the ban on CFCs in order to reduce stratospheric ozone depletion (see Ashford et al. 1985 and Kemp 1997 for discussions). It is clear that this did result in innovation. However, it is not clear that either the rate or direction of such innovation was optimal. Firstly, as with technology-based standards – and unlike market-based measures - the effect is a discrete, once-and-for-all event. Further incentives for future innovations which further reduce adverse environmental impacts are only provided if additional input bans or regulations are introduced in due course. With fewer incentives for firms to innovate, the regulatory authorities are less likely to force such a change since the costs will appear to be considerably higher than under a system where continuous incentives for technological development are in place. Secondly, the ban does not provide any incentives for firms to develop and adopt the most environmentally-beneficial substitute technologies, but merely to discontinue using the banned substance. A tax on ozone-depleting potential (ODP) would not have resulted in as much take-up of HCFC's which, while preferable to CFC's, still have a high ODP.

---

[2] In fact, under very specific conditions, if rules are technology-based, the incentive for the innovator would be greater if the firm is certain that it will generate a rule change. This arises since the innovator's rents are protected by the patent (as under alternative policies) and the market is guaranteed by the rule (unlike under alternative policies). However, this seems unlikely and the innovating firm would face considerable risk in undertaking the necessary investments.

Analogously, different forms of market-based instrument may have different effects on the technological trajectory of the economy. For equivalent environmental targets, auctioned permits and taxes tend to have comparable effects. However, if policy targets are not adjusted in light of increased information then the effects may differ markedly since one is a price-based measure and the other is a quantity-based measure. Thus, Jung et al. (1996) find that when governments pre-commit to a given tax rate or alternatively pre-commit to a given number of permits, the effect under the two regimes will differ since in the case of taxes the "price" of emissions remains constant even as innovation reduces abatement costs.

More generally, work by Pindyck (2000) and others has shown that uncertainties associated with input costs may reduce or delay rates investment, strengthening the case for taxes relative to permits. As price-based instruments taxes may reduce risk from the investors' perspective relative to a quantity-based instrument such as permits (whether grandfathered or auctioned). With reduced risk the rate of investment (and thus innovation) will tend to be higher. However, evidence on the relative importance of this effect in the environmental area is limited.

Kemp (1997) makes the point that it is unreasonable to assume that the target is exogenous even at the point of introduction of the policy. With taxes or auctioned permits the regulator is less likely to introduce stringent environmental policies than under grandfathered permits. As such, the rate of innovation is likely to be lower. While it is certainly true that issues of political economy and rent seeking have been significant in slowing the take-up of emission taxes and auctioned permits, it is important to remember that under either of these schemes distributional and competitiveness concerns can be addressed much more efficiently through other market-based instruments than under grandfathered permit schemes (see Johnstone 1999).

Moreover, unlike under the other systems, with a grandfathered permit system the innovator will face adverse financial effects from reduced permit prices if it is a net seller of permits. Thus, if the innovating firm is a seller of permits, it will have less incentive to allow for the diffusion to other firms (unless the innovation is patented, in which cases incentives will depend upon relative rates of return for permit and technology sales, (see Milliman and Prince 1989)). Significantly, under these restricted conditions grandfathered permits may perform even worse than direct controls such as mandated emission reductions in terms of incentives to induce diffusion. Under direct controls, the only costs are those associated with abatement, but under grandfathering, permit sellers will lose from increased diffusion[3] (see Albrecht 2001).

Despite these qualifications it is generally recognised that in most instances market-based instruments are effective at inducing environmentally-benign research and development, innovation and diffusion (see Albrecht 2001 and Popp 2000 for further discussions). However, empirical analysis is limited. This can be explained by the fact that the "flexible" nature of responses makes it difficult to identify appropriate dependent variables for such an analysis. However, the American Acid Rain programme which introduced tradable permits as a means of reducing SO2 emissions provides some evidence. Under the previous Clean Air Act's rules, firms effectively only had one option for reducing emissions (i.e. to install scrubbers). The allowance trading programme allowed firms more flexibility in their choice of compliance strategies. Indeed, it is significant that very few firms (approximately 10%) complied with the new programme through the use of the technology which had been mandated under the pre-existing regime (scrubbers)[4].

The programme encouraged innovation. On the one hand, there have been improvements in fuel-mixing technologies, allowing firms to shift toward lower-sulphur mixes in a more cost-effective manner. In the late 1980's the theoretical maximum amount of low-sulphur coal that could be mixed with high-sulphur coal was thought to be in the region of 5%, but by the mid 1990's this had risen to 30%-40% (see Burtraw 2000). On the other hand, since the inauguration of the tradable permit system, technological improvements have allowed the price of scrubbers to drop significantly. In 1995, the capital cost of a scrubber sufficient for a 639 MW plant cost less than a scrubber half this size in 1989 (Bohi and Burtaw 1997). This compares with a situation prior to the introduction of the programme in which there had been no appreciable cost-reducing technological developments in flue-gas desulphurisation for 20 years (see Bellas 1998 for an empirical analysis). This compares with a situation prior to the introduction of the programme in which there had been no appreciable

---

3 However, it must again be emphasised that this distinction is not important if innovations are generated by specialist firms which are external to the sector, and thus not themselves involved in the permit market.

4 Although this is certainly at least partly a reflection of the nature of the existing capital stock.

cost-reducing technological developments in flue-gas desul-phurisation for 20 years[5].

Other more formal empirical evidence on the effectiveness of market-based instruments (or other instruments) in inducing innovation is limited. In a study of German firms, Hemmelskamp (1999) finds some support for the positive innovation effects of market-based instruments, particularly for product innovations. Jaffe et al. (2000) report on a study that found that the tradeable permit program used to reduce lead in fuels was very successful at encouraging efficient technology adoption by firms, although the effects on innovation were not explored. They also review a number of studies which find that energy prices have been significant determinants of increased vehicle fuel efficiency. This would indicate that fuel taxes (or permits) would be effective. However, they also refer to a study by Goldberg (1998) that finds that the American "Corporate Average Fuel Economy Standards" (CAFE) have also had positive effects. While often characterised as direct regulation, the programme is perhaps better described as a performance-based firm-level "bubble", in which manufacturers can trade off fuel efficiency improvements for different models within their fleet.

Using a database of German firms, Cleff and Rennings (2000) is one of the few studies to explore empirically the effects of different policy instruments on different types of environmentally-beneficial innovation, including product and process innovations. While they do not find definitive support for the use of one particular instrument, they do find some support for the use of information-based and "soft" instruments such as eco-audits and voluntary approaches.[6] However, work at the OECD (2000) has reached rather different conclusions on the innovation effects of voluntary approaches.

Given the relatively ambiguous nature of the evidence on the effects of different environmental policy instruments on innovation it is important to examine a number of issues associated with such innovation more closely. Moreover, many of these studies use dependent variables which are more closely related to rates rather than directions of innovation. Four issues which are more closely related to the direction of innovation will be addressed in turn: missing markets for certain environmental attributes of in-

---

5 In addition, the costs of transporting low-sulphur coal from the Powder River Basin have fallen, although this is due mainly to institutional, and not technological, factors.

6 Interestingly, they include liability as a soft instrument, while many would consider it to be an economic instrument, albeit one whose price is determined ex post.

novation; technological market failures; joint production of emissions; and, policy incidence. All of these complicate the design of economically environmental policies, particularly when innovation effects are taken into account.

3 Innovation and technological market failures

Environmental market failures are not usually the only failure which affects affect markets in which environmental damages. In the context of this report, there may also be concern about more general innovation-related market failures in environment-intensive sectors. Issues such as capital market failures in research and development, non-excludability and knowledge spillovers, demand-side consumption externalities, credit market failures for potential adopters, and other market failures are pervasive. Thus, with or without the presence of environmental externalities the rate of innovation will be sub-optimal in the absence of government intervention. However, the joint existence of positive technological externalities and negative environmental externalities, may also mean that both the rate and direction of innovation may be inappropriate.

For instance, it is sometimes argued that subsidies should be used to address some of the problems associated with technological market failures. Little work has been undertaken on the evaluation of supply-side environment-related investment subsidies. However, on the demand side, a number of studies have found that subsidies (or tax credits) have been used effectively to support residential energy conservation. Generally such studies find subsidies are very successful in encouraging the rate of diffusion, often much more effective than equivalent tax rates (see Hassett and Metcalf 1995 and Jaffe and Stavins 1995). This might be explained by failures in markets for household credit. However, it is also certainly due in part to the fact that subsidy programmes are not always able to distinguish between households who have been encouraged to undertake the investment because of the subsidy and those who would have undertaken the investment anyway – i.e. there is adverse selection (see Kemp 1997 for a discussion of some of the problems with subsidy programmes).

More importantly the innovation and diffusion which is encouraged may be misdirected. Contrary to the arguments of some, in practice subsidy programmes are not analogous to negative taxes, resulting in similar impacts[7]. While the latter can be designed (but are not always designed) in

[7] Except with respect to entry and exit.

such a way as to be "blind" with respect to the technological solutions adopted, this is rarely the case for subsidy programmes. To one extent or another policymakers are required to determine which investments are eligible for support. As with all such programmes, picking winners is a hazardous exercise.

As noted, information failures are thought to be particularly important technological market failures. Thus, while information-based measures are rarely likely to solve environmental problems by themselves, they can complement other policies very effectively, and may encourage environmentally-beneficial technological change. This is strikingly revealed in a study of product innovations for energy-using household appliances. Looking at the energy-efficiency of air conditioners and water heaters offered for sale in the United States, Newell et al. (1998) estimated the responsiveness of manufacturers to rising energy prices, before and after the introduction of an energy labelling scheme in 1975. The results indicate that the effects of energy price changes on the mean efficiency of appliances supplied by manufacturers rose appreciably (and became statistically significant) once appliances were labelled, encouraging innovation.

Why would this be the case? Assuming that manufacturers were responding to household demand, it is clear that households did not have the information necessary to make informed decisions (or information was too costly to acquire) prior to the introduction of the labelling scheme. Significantly, in a study of high-efficiency lighting in commercial buildings – whose owners would be expected to be better informed than households - Morgenstern and Al-Jurf (1999) also find considerable evidence for the complementary effects of information provision and relative price changes. While such programmes are unlikely to be costless, they are perhaps less likely than subsidies or investment credits to result in a misdirection of innovation.

The internalisation of knowledge spillovers through policy initiatives is a challenging policy task for environmental policymakers. Efforts to encourage such internalisation are very common in other aspects of industrial policy but are not yet common in the environmental sphere. However, measures such as support for research and innovation clusters and networks have been advocated. There is little question that such policies can be an effective complement to more generic policies such as effective intellectual property rights regime and support for basic research. Unlike such measures, however, they seek to direct the pattern of technological change – toward environmentally-beneficial innovation (see Honkasalo 2000 for a discussion of the Finnish experience).

Such programmes may not suffer to the same extent from the problems of "picking winners" associated with subsidies since they are institutional rather than specifically technological in nature. However, under certain conditions, encouraging such co-operation may result in delays in the introduction of environmentally beneficial technologies if there is potential for strategic behaviour. For instance, the US Department of Justice successfully brought an antitrust action against the Automobile Manufacturers' Association partly on the basis of collusion in delays of the announcement of process innovations which would have reduced environmental impacts (see Hackett 1995).

4 Missing markets for environment-related attributes and product innovation

Closely related to the issue of technological market failures is the issue of missing markets for environmentally-relevant product attributes. Many studies on the innovation effects of environmental policy instruments implicitly assume that the only market which is missing is that for the environmental externality. However, in many cases this is not the case. The example of post-consumption solid household waste is instructive. A study by Eichner and Runkel (2000) shows that if there are not "indirect markets" for product toxicity, then the environmental attributes of products will be sub-optimal - i.e. firms will underinvest in the development of products which are not toxic. Similarly if there are not "indirect markets" for recyclability then firms will underinvest in the development of products with attributes which make this more feasible. Hence, this analysis shows that besides environmental externalities there are further sources for inefficiency, namely missing markets for product design.

The key point is that even if environmental policies - such as technological standards for landfills or incinerators - are targeted at waste at the post-consumption phase, this will not result in improved product design. Under such measures regulators are unable to target products differentially, and thus manufacturers and product designers face little incentive at the individual level to incorporate such elements in their products. Even if such measures are financed through volume-based waste fees – as is increasingly the case – the transmission of signals back to product manufacturers will be blunted by the mixed nature of municipal solid waste streams.

Thus, incentives for firms are often inadequate. Such "technical externalities" are pervasive in the markets for used appliances and parts of appliances, as well as packaging waste. For instance, upstream product designers and manufacturers are not encouraged to design for recycling since downstream users may not face appropriate financial incentives to purchase products which are recyclable. The end result is that the benefits of particular types of design may be less than the costs, but there is no way for this information (and appropriate incentives) to be transmitted to product designers. Technological innovation with respect to product design will be misdirected, even in the presence of measures usually proposed by economists such as volume-based solid waste charges.

Thus it may be necessary to introduce complementary policies to address the issue of "missing markets", which can encourage the design and development of environmentally-preferable products. In most cases it will not be possible to create the market directly through policy interventions. However, there may be other remedies. For instance, in order to encourage improved product design, measures such as deposit-refund systems or product take-back programmes may be effective since they can "bracket" the missing market and transmit signals back to designers and manufacturers. Directed government support for "Green Design" has also risen up the policy agenda in many OECD countries, and consumer durables have been a primary area of focus. Unfortunately, all such measures impose significant information requirements on policymakers and in some cases administrative costs for public authorities and private firms and households.

5 Point of policy incidence and innovation

To a certain extent, the focus on the importance of instrument choice when evaluating the innovation effects of environmental policy may be misplaced. In some senses it might be more important to hit in the right place in the product cycle, rather than to do so with the right stick (or carrot). Indeed, it is frequently assumed that it is possible for the policymaker to target the environmental externality directly. However, this is very rarely the case. Arguably, a CO2 tax is the only existing example. Due to the high administrative costs or even technological infeasibility in other areas, almost all policies are targeted at some proxy for the damage rather than the damage itself.

For instance, taking the example of acidification from SO2 emissions emitted by the electricity supply industry, policies could target any of points listed in table 1. Generally speaking, there is a trade-off between the accuracy of targeting the externality and the administrative costs of doing so as you shift down the list. The administrative savings from not targeting damages are directly are well-understood. However, the cost of shifting away from targeting the externality are not well understood.

Table 1. Environmental impacts and policy incidence

Target	Point of incidence	Example of direct regulation	Example of market-based instrument
Environmental damages	Critical loads	Restricted entry in non-attainment areas	Permits denominated w.r.t. critical loads
Environmental pressures	SO2 emissions	Performance standard for emission levels	Emissions permit
Material inputs	Coal	Restriction on use of high-sulphur coal	Tax on sulphur content of fuel
Production/combustion process	Combustion/abatement technology	Mandated use of scrubbers	Accelerated depreciation for scrubbers
Products	Electricity	Restricted access to the electricity grid	"Green" electricity tax exemption

In the short-run the costs of "missing the target" may be minimal since proxies will always be chosen which are highly correlated with the ultimate damage. However, the long-run effects may be considerable since the policy will, in effect, encourage firms and households to save on the proxy and not on the damage. Moreover, the relationship will necessarily become weaker through time if there is any degree of substitution between the proxy and the ultimate environmental impact.

For instance, in many European countries vehicle ownership taxes are differentiated according to vehicle weight or engine size in an effort to reduce local and global air pollutants. While there is a relatively strong correlation between emission levels and vehicle weight, if the measure is significant enough to encourage vehicle redesign, this relationship becomes weaker through time. By trying to save on the characteristic which is taxed, manufacturers will be unconstrained with respect to emission levels. Similar issues arise in the area of agriculture, where "proxies" are used extensively due to the high administrative costs of target non-point source pollutants directly.

The importance of this issue has not been examined empirically, even though Sandmo (1976) raised the issue three decades ago. However, in recent theoretical papers on the issue of incidence both Schmutzler and Goulder (1997) and Fullerton et al (1999), look at the welfare effects of output taxes relative to emission taxes. Not surprisingly, they find that the welfare costs of the former can be much greater than the latter. Dinan

(1993) looks at the example of the application of a proxy – targeting waste externalities through a virgin materials tax – and reaches similar conclusions. Work on input taxes might be more interesting, given the prevalence of their use as proxies in many elements of environmental policy and given the rather different implications that they have for innovation and factor substitution.

However, what such studies do show is the importance to distinguish between instrument choice and the point of incidence when assessing the innovation effects of environmental policy. The case for economic instruments relative to direct forms of regulation is often made under the assumption that they both target externalities equally accurately. However, two points (point of incidence and instrument choice) are being conflated. For instance, a performance standard based on emission levels is likely to target the externality more directly than a tax in which the tax base is related to an input. While the latter may result in greater static allocation efficiency, it may result in misdirected pattern of innovation. The examples in table 2 make this distinction clear.

6 Innovation when pollutants are joint-products

Pollution emissions are best understood as joint-products, not just with respected to commodities, but also with respect to each other. Firstly, emissions of different pollutants are often highly correlated. table 2 presents correlation coefficients calculated for four different pollutants, based upon almost 3,000 observations from the US EPA's vehicle emissions database.

Table 2. Correlation between different vehicle emissions

		THC	CO	NOX
CO	Pearson Correlation	.828		
	Sig. (2-tailed)	(.000)		
NOX	Pearson Correlation	.132	.107	
	Sig. (2-tailed)	(.000)	(.000)	
CO2	Pearson Correlation	.026	-.002	.265
	Sig. (2-tailed)	(.171)	(.920)	(.000)

Correlation is significant at the 0.01 level (2-tailed). 2,851 observations

One pair of pollutants (CO and THC) have particularly a high positive correlation. This can be seen visually in figure 1 which provides a scatter plot for observations between 1990 and 1995. Interestingly only CO2 and CO, have a negative relationship - but it is not significantly different from zero.

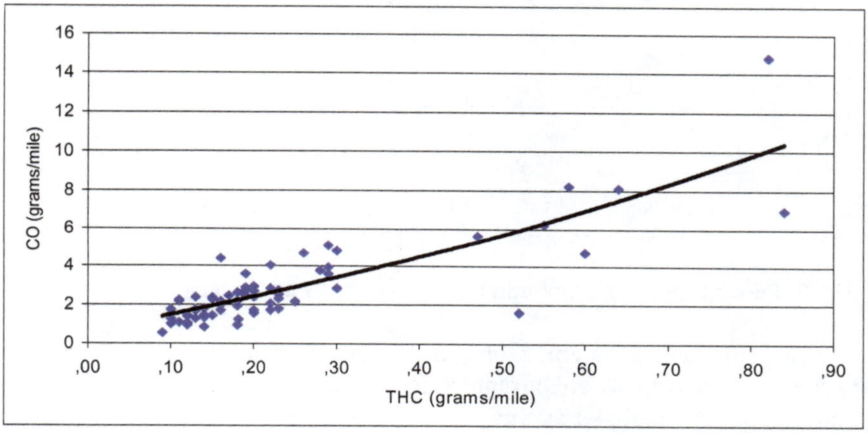

Fig. 1. Scatter plot of CO vs. THC emissions

The jointness of emissions is partly a consequence of the nature of the production or combustion process. However, and perhaps more significantly in policy terms, different emissions are often jointly reduced through abatement. The close link between CO and THC emissions illustrated in figure 1 is partly a consequence of the application of end-of-pipe catalytic converters. In effect, the degree of 'jointness' is endogenous to the policy measure.

However, in recent years there has been a marked shift toward the use of changes in production processes rather than end-of-pipe abatement. (See figure 2 for some data derived from the American Census of Manufactures.) Production and abatement are no longer separable. This necessarily results in "bundling" of emissions associated with different technologies. In effect, the shift toward abatement through changes in products and production processes is likely to lead to economies of scope across different types of pollutant. Unfortunately, the importance of this has not yet been explored systematically.

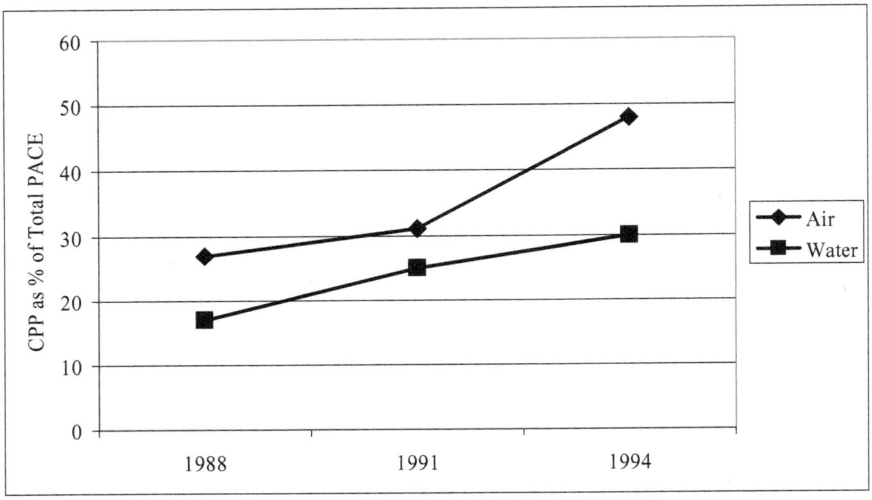

Fig. 2. Percentage of total pollution abatement costs attributed

The joint nature of emissions, may appear to simplify the regulator's task when seeking to encourage innovation. Ancillary or complementary benefits will be realised as the overall level of environmental performance rises. However, in some cases they are substitutes. For instance, it has long been recognised there is significant potential for "shifting" of environmental burdens between different types of emissions and even between media. Depending upon substitution possibilities in production and consumption constraining emissions may result in increased emissions of another sort. Grafton and Devlin (1994) explore the effects of regulating one emission when another (substitute) emission is left unconstrained. In the long term as firms will innovate in a manner which results in higher emissions of the latter, potentially resulting in decreases in overall environmental quality[8].

Once again, motor vehicles provide interesting examples of potential substitution. For instance, measures to improve fuel-efficiency (and thus reduce CO2) lead to higher combustion temperatures and thus higher NOx emissions. To find an optimum for controlling pollutants the trade-off between CO2 and NOx controls had to be managed. When catalytic converters were introduced this problem disappeared, as overall reductions of more than 80% of all pollutants and air toxic were achieved. However, there was a slight reduction of fuel-efficiency (and thus increase in CO2

[8] In some cases it is quite likely that emissions will be complements in the short-run, but substitutes in the long-run.

emissions) of a few percentage points. In addition, nitrous oxide (N_2O) emissions were 3 to 5 times higher compared to vehicles without catalysts[9]. Khazoom (1996) looked at causality in the opposite direction (i.e. from CO_2 to local air pollutants) and pointed out that while the American Corporate Average Fuel Economy standards may have led to fuel-saving (and carbon-saving) technological development, it may have had negative consequences for local environmental quality.

7 Implications for the choice of indicators for environmental innovation

Assessing whether or not different environmental policy instruments have positive impacts on technological innovation is dependent upon the existence of appropriate indicators. However, given the discussion above it is clear that the optimal direction of innovation can be complicated by the existence of various factors such as the existence of market failures, the joint production of emissions and abatement, the presence of missing markets, and the targeting of policies at some remove from the externality itself. As such, the choice of indicator is an important but hazardous exercise.

In the OECD framework, indicators have been developed within the 'pressure-state-response' framework (see OECD 2001). Indicators for environmental innovation are, almost by definition, 'response' indicators – reflecting adjustments within the economy to environmental conditions. However, in some cases it may be possible to derive pressure indicators. For instance, much of the work which has been carried out in the context of 'decoupling of environmental pressures from economic growth' (OECD 2002) can be understood as reflecting, at least in part, innovation. However, at the macroeconomic level, they also reflect other factors such as changes in sectoral composition of the economy.

Therefore, it is important to identify what we might consider to be possible indicators specifically for environmental innovation. A partial list might include the following:

- Patents for innovations which result in improved environmental performance;
- Percentage of research and development which is related to environmental matters;

---

9 However, it is worth noting that this increase is of minor importance, as transport contributes only 3% to total N_2O emissions (primarily from agriculture), and is less than 1% of total greenhouse gas emissions (ECMT, 2001).

- Adoption/diffusion rates for environmentally-benign technologies and product types;
- Normalised emission rates for particular production processes or product types; and,
- Investment in product designs which reduce resource use in production and use.

Assessing such indicators requires careful analysis. According to the OECD (2001), indicators for environmental issues should fulfil the following criteria:

- Policy relevance and utility for users
    - Provide a representative picture of environmental conditions, pressures or responses;
    - Be simple, easy to interpret, and able to show trends over time;
    - Be responsive to changes in the environment and related human activities;
    - Provide a bases for international comparison;
    - Be either national in scope or applicable to regional issues of national significance; and,
    - Have a threshold or reference value to allow for ease of interpretation.
- Analytical soundness
    - Be theoretically well-founded in technical and scientific terms;
    - Be based on international standards and international consensus about its validity; and,
    - Lend itself to being linked to economic models, forecasting and information systems
- Measurability
    - Be readily available or made available at reasonable cost;
    - Be adequately and of known quality; and,
    - Be updated at regularly in accordance with reliable procedures.

Very few, if any, of the general environmental indicators listed in the report, satisfy all of these criteria. Applying these same criteria for the specific case of innovation-related indicators in the environmental is likely to be even more problematic. However, rather than seeking to identify the extent to which individual indicators satisfy all criteria, particular attention will be paid to those criteria which relate specifically to analytical soundness.

In particular, it is important to recognise the implications of the discussion above, which highlighted the distinction between measures of the 'rate' of innovation, and 'direction' of innovation. Since all response and pressure indicators of the sort enumerated above are at least one step removed from the ultimate policy objective (the state of environmental conditions), it is important that the indicators seek to reflect the 'direction' of innovation in a manner which is useful for assessment. There can be many slips 'twixt the cup and the lip'.

Therefore, a successful indicator must reflect not only the rate of innovation but also the direction of innovation. How well do the possible indicators listed above capture these two effects? Problems in accurately measuring 'rate' of innovation are inherent in any measure which does not reflect the efficacy of measures of the sort discussed above. There is, for instance, a lively debate as to whether public investment in research and development 'crowds in' or 'crowds' out private investment in research development.

Thus, indicators such as the percentage of research and development or levels of investment in 'environmental' areas (whether defined by sectoral or commodity classification) are only useful insofar as they are examined jointly with other indicators which reflect the efficacy of such expenditures. However, perhaps even more fundamental is the assessment of the analytical soundness of different indicators with respect to the direction of innovation.

For example, if we take the case of patents for innovations which result in improved environmental performance, the direction of innovation is dependent upon the choice of technologies which are considered 'environmental' in nature. Arguably, such technologies are easier to identify for technologies related to 'end-of-pipe' abatement than changes in production processes, and as such are likely to be disproportionately reflected in the measurement of the indicator. However, since changes in production processes are often more economic in the longer-run, an increase in the indicator might not reflect improved environmental innovation.

Similarly, if we take the case of waste-related innovations in the area of product design as another example, it might well be easier to develop indicators which relate to ease of recycling than those which relate to waste prevention. Since some innovations which result in improved recycling rates for material inputs may substitute for overall waste prevention, an increase in particular indicator may not reflect an optimal direction of innovation.

It is, therefore, quite possible to develop indicators which appear to reflect environmental innovation, but the distance between such indicators and ultimate environmental conditions is such that the relationship between the two can not be taken as given. Such problems are likely to be particularly problematic for indicators at the macroeconomic or sectoral level. Assessing environmental 'innovativeness' at such an aggregate level is an exercise which needs to be undertaken with great care, for all of the reasons discussed above.

8 Conclusion

The first lesson to be drawn from this study is the importance of not confusing the optimal rate of innovation with the optimal direction of innovation. Given that the latter may be more difficult to achieve than the former, this point is far from trivial. Ascertaining what direction is optimal is, of course, endlessly problematic. However, in abstract terms it should be that path which is cost-minimising in the long-run with respect to the realisation of the given environmental objective.

The second lesson is that in order to ensure that there is no misdirection of innovation, policies should be targeted at the ultimate environmental damage as closely as is administratively feasible. This is easier said than done, and direct targeting is rarely possible - except with the notable exception of carbon dioxide. However, the costs of mistargeting, particularly in the long-run, need to be more widely-recognised than is usually the case at present.

Moreover, targeting must be undertaken in an "integrated" manner. If different environmental objectives are narrowly defined in terms of objectives and targeted sequentially (as is usually the case) then economies may be pushed onto a technological trajectory which is relatively inefficient, and potentially environmentally perverse. Realisation of one environmental objective today may come at the expense of the realisation of other objectives in future years.

The third lesson is that it is important to look at the structure of markets, technological factors associated with production and abatement, and the precise nature of the environmental damage to be mitigated, when designing policies to encourage environmentally-beneficial innovation. Blanket prescriptions in favour of one instrument over another are not reliable. However, those instruments which allow for flexibility in implementation and provide continuous incentives for innovation should be used wherever possible.

9 References

Albrecht J (2001) Tradable CO2 Permits for Cars and Trucks. Journal of Cleaner Production 9:179-189

Ashford NA, Ayers C, Stone R (1985) Using Regulation to Change the Market for Innovation. Harvard Environmental Law Review 9(2): 419-466

Bellas AS (1998) Empirical Evidence of Advances in Scrubber Technology. Resource and Energy Economics 20: 327-343

Bohi DR, Burtaw D (1997) SO2 Allowance Trading: How Experience and Expectations Measure Up. Washington D.C.: Resources for the Future Discussion Paper 97-24.

Burtraw D (2000) Innovation Under the Tradeable Sulphur Dioxide Emission Permits Programme in the US Electricity Sector. In OECD (ed) Innovation and the Environment. Paris

Cleff T, Rennings K (2000) Determinants of Environmental Product and Process Innovation - Evidence from the Mannheim Innovation Panel and a Follow-Up Telephone Survey. In Hemmelskamp J, Rennings K, Leone F (eds) Innovation-Oriented Environmental Regulation. Heidelberg

Grafton RQ, Develin RA (1994) Les Permis d'émission et les charges: efficacité et substitutabilité. L'Actualité Economique: Revue d' Analyse Economique 70(2): 159-176

Dinan T (1993) Economic Efficiency Effects of Alternative Policies for Reducing Waste Disposal. Journal of Environmental Economics and Management 25(3): 242-256

ECMT (European Conference of Ministers of Transport) (2001) Vehicles Emission Reduction. Paris

Eichner T, Runkel M (2000) Efficient and Sustainable Management of Product Durability and Recyclability. Working Paper. Department of Economics University of Siegen

Environmental Law Institute (1998) Barriers to Environmental Technology Innovation and Use. Washington D.C.

Fullerton D, Hong I, Metcalf GE (1999) A Tax on Output of the Polluting Industry is not a Tax on Pollution: The Importance of Hitting the Target. Paper Presented at the NBER-FEEM Conference on Distributional and Behavioural Impacts of Environmental Policy in Milan

Goldberg P (1998) The Effects of the Corporate Average Fuel Efficiency Standards in the U.S.. Journal of Industrial Economics 46: 1-3

Hackett SC (1995) Pollution-Controlling Innovation in Oligopolistic Industries: Some Comparisons between Patent Races and Research Joint Ventures. Journal of Environmental Economics and Management, Volume 29, Issue 3 , Part 1, Pages 339-356

Hasset KA, Metcalf GE (1995). Energy Tax Credits and Residential Conservation Investment. Journal of Public Economics 57: 201-217

Hemmelskamp J (1999) The Influence of Environmental Policy on Innovative Behaviour: An Econometric Study. IPTS Working Paper. Seville

Henriques I, Sadorsky P (1996) The Determinants of an Environmentally Responsive Firm: An Empirical Approach. Journal of Environmental Economics and Management 30: 381-395

Honkasalo A (2000) Eco-efficiency, Entrepreneurship and Co-operation: The Finnish Environmental Cluster Research Programme. In OECD (ed) Innovation and the Environment. Paris

Jaffe AB, Stavins RN (1995) Dynamic Incentives of Environmental Regulations: The Effects of Alternative Policy Instruments on Technology Diffusion. Journal of Environmental Economics and Management 29: 43-63

Jaffe A, Newell R, Stavins R (2000) Technological Change and the Environment. Resources for the Future - Discussion Paper 00-47

Johnstone N (1999) Tradable Permit Systems and Industrial Competitiveness: A Review of Issues and Evidence. In OECD (ed) Domestic Tradable Permit Systems for Environmental Management: Issues and Challenges. Paris

Johnstone N (2001) Recycled materials study

Jung C, Krutilla K, Boyd R (1996) Incentives for Advanced Pollution Abatement Technology at the Industry Level. Journal of Environmental Economics and Management 30: 95-111

Kemp R (2000) Technology and Environmental Policy: Innovation Effects of Past Policies and Suggestions for Improvement. In OECD (ed) Innovation and the Environment. Paris

Kemp R (1997) Environmental Policy and Technological Change. Cheltenham

Khazoom JD (1996) An Econometric Model of the Regulated Emissions for Fuel-Efficient New Vehicles. Journal of Environmental Economics and Management 28: 190-204

Kneese A, Schultze C (1975) Pollution, Prices and Public Policy. Washington

Milliman SR, Prince R (1989) Firm Incentives to Promote Technological Change in Pollution Control. Journal of Environmental Economics and Management 17: 247-265

Morgenstern RD, Al-Jurf S (1999) Can Free Information Really Accelerate Technology Diffusion. Technological Forecasting and Social Change 61: 13-24

Nentjes A, Wiersman D (1987) Innovation and Pollution Control. International Journal of Social Economics 15: 51-71

Newell RG, Jaffe A, Stavins R (1998). "The Induced Innovation Hypothesis and Energy-Saving Technological Change" Quarterly Journal of Economics, Vol. 114, No. 3, August 1999, pages 941-75.

OECD (2002) Indicators to Measure Decoupling of Environmental Pressure from Economic Growth. SG/SD 1/FINAL

OECD (2001) Ten Environmental Indicators. ENV/EPOC 11/ADD1

OECD (2000) Voluntary Approaches for Environmental Policy: An Assessment. Paris

Pindyck RS (2000) Irreversibilities and the Timing of Environmental Policy. Resource and Energy Economics 22(3): 233-259

Popp D (2000) Induced Innovation and Energy Prices. NBER Working Paper 8284. Cambridge

Sandmo A (1976) Direct versus Indirect Pigovian Taxation. European Economic Review 7(4) 337-349

Schmutzler A, Goulder LH (1997) The Choice between Emission Taxes and Output Taxes under Imperfect Monitoring. Journal of Environmental Economics and Management 32: 51-64

United States Department of Commerce, Economics and Statistics Administration (various years) Current Industrial Reports: Pollution Abatement Costs and Expenditures. Washington D.C.

Indicators for a Sustainable Technology Development – A Dynamic Perspective[1]

Christian Sartorius

1 Introduction

From the more technical perspective, the discussion of sustainable innovations is basically concerned with two questions. What is the underlying conception of sustainability and how do the innovations in question conform to the chosen conception? In this context, indicators of sustainable innovations primarily deal with questions of operability and comparability (see e.g. Pearce et al. 1989 and Rennings 2000 for an overview).

Inclusion of the economic perspective then leads to the question whether and under which conditions a sustainable innovation will also be marketable. Are its properties acceptable for the potential customers and can it be produced at an acceptable price? Once the innovation meets the conditions for successful market entrance, also its macroeconomic impact, particularly its welfare, employment and, possibly, social distribution effects, will be of interest.

Once it turns out that an innovation shows promising ecological and social properties but at least temporarily lacks economic competitiveness, it is a possible role of the state to support this technology until it can successfully compete with its less sustainable counterpart. Since the intervention of the state usually takes the form of a market regulation, the next question typically asks which instruments for such political interventions exist and which ones appear to be most suitable. Eventually it may turn out that even a mix of regulative measures is needed to properly account for the complexity of circumstances in which the innovation arises (Klemmer et al. 1999).

While, up to this point, the discussion of sustainable innovations has already reached a considerable degree of sophistication, one major point is still missing. Although, in the context of regulatory instrument mixes, the diversity and complexity of circumstances is well acknowledged, time as

---

[1] Funding of this research by the German Federal Ministry for Education and Research (grant 07RIW5C) is gratefully acknowledged. I thank my colleagues Jan Nill and Stefan Zundel for valuable comments to earlier versions of this paper.

an important factor of influence is neither explicitly mentioned nor, all the less, systematically investigated. In fact, the neglect of time is a major omission because the circumstances change with time and the respectively most appropriate regulatory measures with them. Since each instrument causes to the state specific costs, it should also be clear that the necessary expenses will vary considerably with the changing circumstances and, of course, with time.

In this paper, time will be accounted for more thoroughly. In particular, it is assumed that along with the change in circumstances, periods of stability (where establishing a different technological regime requires much effort) alternate with periods of instability (where such a shift is more easily achieved). It is further assumed that in the search for the lowest possible cost of implementing an innovation, it is possible to identify and even strategically use the latter phases of instability. After a short discussion of the relevant concepts of sustainability and sustainable innovations in section 2, it will be shown in section 3 that the alternation between stability and instability exists and how it may be used to achieve better long-term sustainability. In order to account for this dynamic conception of sustainability, a broad set of relevant factors and the corresponding indicators will be developed and a proposal for their integration made in section 4. In section 5, the operability of this set of indicators will be illustrated in the light of a series of innovations following the phase-out of ozone-depleting CFCs in the 1980s and 1990s. Finally, section 6 will conclude.

2 Sustainability and its assessment

Sustainability is usually discussed as a state or, better, a development in which three kinds of (conflicts of) interests are met (or resolved) simultaneously: (i) the interest of the present generation to generally improve their actual living conditions (i.e. economic sustainability), (ii) the search for an equalisation of the living conditions between rich and poor (i.e. social sustainability), and (iii) the interests of future generations that are not to be compromised by the actual need satisfaction of the present generation (i.e. ecological sustainability). It is intuitively clear that particularly less developed countries show a stronger tendency to consider the (over)use of the environment as one of their more important potentials for earning a sufficient income and that therefore a very unequal distribution of resources is one of the major causes for environmental destruction. Since this issue is subject to intense political discussion and continued negotiations between most countries, the normative character of social (re-)distribution is readily accepted as an argument to exclude it from the scientific discourse. Although balancing the interests of succeeding generations is a normative issue as well, the lacking possibility of the future generations to participate in the corresponding political discussion is in this case taken as a justification and as a potential for science to make fruitful contributions. Consequently, the discussion of sustainability particularly among economists essentially focuses on the question how to allow for the strongest possible growth now without compromising the potential for growth to persist in the future.

2.1 Weak vs. strong sustainability

The main precondition for such equal treatment of successive generations is the preservation of a pool of natural resources and man-made capital that provides each generation with identical starting conditions, that is, with the opportunity to have its activities based on equivalent sets of man-made and natural capital. This conceptualisation of sustainable development as "non-declining wealth" (Pearce et al. 1989) finds two basically different expressions. On the one hand, economists in the tradition of Hartwick (1978) and Solow (1986) argue that a society using an exhaustible stock of resources could enjoy a constant stream of consumption over time if it invested all the rents from tapping on those resources, that is, if it held the overall capital stock constant. Evidently, this weak approach to sustainability is based on the implicit assumption that both natural and man-made capital are complete substitutes. While this assumption may be met in some cases, it does not hold in general. For many types of natural assets (e.g. an endangered species, a habitat or the ozone layer) technical substitutes do not exist. In general, the latter argument applies even more to the capability of the natural environment to assimilate the by-products of human activities than to its function as a mere supplier of input resources. It is for this reason that the central role of substitutability between man-made and natural capital is essentially questioned by the supporters of the concept of strong sustainability. According to the so-called 'management rules' (Daly 1990), for instance, proponents of the latter concept claim that (i) the harvest rates of renewable resources is not allowed to exceed their rate of regeneration, (ii) the rates of generation of by-products from the production, use, and disposal of goods should not exceed the respective assimilation rates of the ecosystem, and (iii) the exploitation of exhaustible resources has to be compensated through replacement with equivalent (renewable) alternatives. So, substitutability has to be proven rather than simply being assumed. With regard to the properties qualifying a technology as sustainable, the requirements in a context of strong sustainability are evidently much stronger than in a context of weak sustainability.

2.2 Sustainability indicators

The relation between weak and strong sustainability is also mirrored in the indicators used for their operationalisation. According to the weak concept, the development of a given economy is considered sustainable, if the total savings are higher than the combined depreciation of both, natural and man-made capital. Since the net investment into man-made capital and the damage to the environment are both measured (e.g. by green GDP accounting) and freely aggregated in terms of money, they are evidently treated like full substitutes (Pearce and Atkinson 1993).

Unlike weak sustainability, concepts of strong sustainability specify the natural capital in terms of its physical function rather than the costs of actual damage caused to it. The logic of this approach is based on the assumption that in order to continue to rely on certain essential functions of the environment (e.g. assimilation of waste or supply with resources), the ecosystem or at least certain parts of it have to be kept intact. Although this approach does not exclude monetisation in principle (e.g. in terms of the opportunity costs of the avoided or restricted use of the environment), the (however aggregated) monetary figure does not suffice to eventually specify the state of sustainability. Instead, it is necessary to follow the following three-step procedure and to (i) identify those elements of the natural capital that are essential for the maintenance of the ecosystem's stability or resilience, (ii) select those elements that are related to, and possibly endangered by, economic activities, and (iii) derive a set of indicators each of which reflects the actual condition of a specific aspect of the environment and puts it into relation to the sustainable state as determined by any suitable management rule (see Opschoor and Reijnders 1991).

Typical examples of the latter approach are Pressure-State-Response (PSR) indicators like the one employed by the OECD. Here, the causes of environmental problems ("pressure"), the actual state of the environment ("state"), and efforts to solve the problem ("response") are monitored and quantified in separate modules. Problems however exist with the assignment of counter-measures ('response') to specific pressures and states. While it is possible in the short run to quantify the effect of the latter measures in terms of a reduction of those processes or their side-effects that caused the corresponding pressure in the first place, many counter-measures later turn out to be themselves not without side-effects such that the relaxation of pressure in their target field may go along with the increase of pressures in other fields. This kind of uncertainty is characteristic not only for environmental innovations.

2.3 Critical loads and non-linearity

While a PSR-like indicator represents a first important step to the assessment of the causes and development of environmental problems, it supposes a correlation between pressure and response that is misleading for the following reason (Rennings and Wiggering 1997): The logic underlying the PSR approach implies that stronger (weaker) efforts to counteract an environmental problem by means of the best-available technology will generally lead to the alleviation (enhancement) of the pressure and, thus, to the improvement (deterioration) of the condition of the environment. Unfortunately, with regard to the environment, such a "linear" relation between causes and effects is not the rule. In contrast, effects like the following are frequently observed. Although in a certain agriculturally dominated region the intense use of mineral fertilisers was common practice for quite a while, contamination of the ground-water with nitrate could be observed only recently – with a strongly increasing rate. Due to the existence (and transgression) of carrying capacities or buffer capacities, such non-linear processes typically show sudden changes or even jumps. Returning to a sustainable state then not only requires the reduction of emissions below the respective critical load or critical level. Since the latter may itself be adversely affected by the harm, it additionally requires the repair of the damages that had so far been caused by the excess emissions.

3 Sustainable innovations in evolutionary perspective

It should have become evident at this point that innovations can usefully be integrated only into a concept of strong sustainability. Weak sustainability, by contrast, does not only fail to question the crucial substitution between natural and man-made capital; it also fails to differentiate which kind of technology or innovation is employed and whether innovative activities are shown at all.

But despite their significance in the context of sustainability, innovations also play an ambiguous role. On the one hand, they offer a potential to redress sustainability once it is lost; on the other hand, they are often also the cause for just this loss. After discussing some basic properties of sustainable innovations, the major part of this section will focus on two aspects of innovations that are extensively discussed in evolutionary economics: uncertainty and path dependence.

3.1 Innovations and sustainability

The usual (economic) understanding of the process of innovation is documented in the 'Oslo Manual' of the OECD (1997) and essentially distinguishes between process, product, and organisational innovation. While a process innovation basically refers to the (quantitative) relation between input factors and output commodities and a product innovation typically comprises a change in the (qualitative) properties of the output, organisational innovations can be associated with both, qualitative and quantitative changes (Rennings 2000). In all three cases, the term innovation refers to efficiency increases, that is, to changes in the production of goods and services that ultimately allow for a better satisfaction of certain needs and desires of the consumers with the same set of input factors or, equivalently, for the satisfaction of the same needs and desires with less input.

Sustainable innovations could basically be defined in the same way as ordinary innovations, however with the important restriction that the efficiency increase is not allowed to violate the chosen sustainability (e.g. Daly's management) rules. However, since it is evident that the current human way of life leads to transgressions of the sustainability boundary in many and profound ways and that the existing institutions (including codified rules, customs, habits, and social preferences) are broadly coherent with just this lifestyle, efficiency changes under the proviso of sustainability may sometimes be achieved more readily through institutional or social than through technical innovations. In the context of sustainability, it is therefore necessary to broaden the view from the merely technical towards the social and political aspects of innovations. In accord with these thoughts, Klemmer et al. (1999; see also Rennings 2000) broadly define the term 'environmental innovation' as all measures of relevant actors that lead to the development and application of new ideas, behaviour, products and processes and, thereby, contribute to a reduction of environmental burdens or to ecologically specified sustainability targets. This may include process and product innovations, organisational changes in the management of firms, and, on the social and political level, changes in environmentally counter-productive regulation and legislature, consumer behaviour, or lifestyle in general. This emphasis on social innovations is all the more important because unsustainable development itself is often the result of "technology outpacing changes in social organisation" (Norgaard 1994). Moreover, after an intense and extended discussion in environmental economics about the "right" instruments towards an environmentally sound, sustainable development, it more and more turns out that there is not a single suitable instrument. Instead, it seems to depend on the respective circumstances (e.g. the type of competition or information asymmetries), whether Pigovian taxes, markets for pollution rights, the setting of standards or even temporary subsidisation of promising innovations is the more effective instrument (Rennings 2000). Jaenicke (1999) even goes one step further by claiming that the relevance of instruments for environmental policy has generally been overemphasised. Instead, the discussion should focus on other elements of a successful environmental policy such as long-term goals, mixes of instruments, policy styles, and constellations of actors.

Altogether, the above emphasis on social and political aspects makes clear that the success of sustainable innovations depends on more than their mere technical (or even economic) superiority. This is all the more evident when, according to the following suggestions, sustainability is considered as the property of an entire system rather than being associated with a specific innovation.

3.2 Fundamental uncertainty

It is the wide variety and high complexity of interactions between human actors and between the latter and their natural environment that renders human (economic) activities as well as their environmental effects highly unpredictable particularly in the long run. However, the uncertainty accruing in this context is not just a matter of probability distributions within a known or assumed set of possibilities. Instead uncertainty is better characterised as ignorance in the face of novel, fundamentally unpredictable, events. So the question arises how to deal with this fundamental uncertainty. If complete knowledge about the set of available alternatives is lacking, actors cannot maximise the expected utility of alternative choices and, thus, rational decisions cannot be made. Moreover, rational choice theory assuming fixed sets of individual preferences that basically include all possible alternatives may simply turn out to be underdetermined in the face of real novelty.

Therefore, it may be advisable to look at the solution of (long-run) problems related to fundamental uncertainty and inflexible preferences from a completely different perspective: the one represented by Darwin's approach to evolution in nature. Like society, nature is characterised by the complex interaction between its constituents, the living organisms and their physical environment, and thus by the existence of fundamental uncertainty and non-linearity which together give rise to the formation of new species or the sudden extinction of major parts of the existing biosphere. In order to "manage" such unpredictable processes, nature relies on the principles of random variation and natural selection – with diversity created by random mutation and recombination within the existing genetic pool and selection resulting from continuous competition of species for a limited set of resources.

A further step toward an increased problem solving capability in nature and, ultimately, in man is based on the capability of an organism to undergo specific or individual adaptation to varying circumstances and to transmit the acquired knowledge to other organisms – that is to learn and communicate. While evolution on this level is based on social norms, individual values, and ideas rather than material genes, the basic principles nevertheless remain essentially unchanged (Sartorius 2003, especially ch. 4). Initially, the perception of a problem leads to the assessment of a variety of alternative approaches to its solution. Those approaches giving rise to a solution of the problem are selected; those that fail are rejected. The solutions with the best performance are further modified and tested in subsequent rounds of selection. The wider the variety of alternative approaches the higher is the probability that at least one of them may perform better than in the status quo. With respect to human behaviour, special use of evolutionary principles has been made by many proponents of evolutionary economics: Schumpeter (1934), for instance, emphasises the relevance of entrepreneurial creativity as a source of new problem solutions; Hayek (1978) interprets market competition as a process of selection (and detection) of superior goods by means of the willingness-to-pay on the demand side; and Nelson and Winter (1982) show how profit may serve as the selecting force that leads to the persistence of some innovations and to the vanishing of most others. A particular case of evolution leading to the solution of unprecedented problems is the selection of co-operation rules on the group level, a task that could never be fulfilled by individuals on the basis of their mere rationality (Hayek 1978; Sartorius 2002). In this context, (environmental) sustainability can indeed be interpreted as co-operation (i.e. as an expression of fair behaviour) between succeeding generations.

The relevance of fundamental uncertainty and the corresponding problem solving capability for sustainability is quite evident. Human activities frequently generate adverse environmental side-effects which, due to the complexity of their interaction with the environment, are often unforeseen. In the search for (long-term) sustainability indicators, it therefore makes little sense to exclusively rely on indicators that are related to specific environmental problems and their causing agents since they may be subject to considerable variation over time. This does not at all imply that the determination of critical substances and the application of critical thresholds do not make sense. Especially in the short run they are even indispensable. However, in the long run, that is in the time perspective in which the sustainability concept is usefully applied, an indicator for sustainability also has to account for the conditions under which the identification of problems as well as the search for the corresponding solutions and their translation into the appropriate measures takes place. Rather than referring to specific innovations whose characterisation as being sustainable can only be a temporary one, sustainability being the property of a system should be determined with reference to the system's general capability to bring about a variety of potentially useful innovations and, should the occasion arise, to allow for the ready implementation of the most promising alternative. In short, sustainability also, and from the evolutionary perspective predominantly, includes the flexibility and versatility of the entire system to allow for a quick and effective response to whichever environmental problems arise (see Erdmann 2000).

3.3 Irreversibility and path dependence

Beside fundamental uncertainty and the need for diversity following from the preceding argument, the complexity of multiple-interaction systems has another at least equally important consequence for the sustainability discussion. If the sequence of events within a complex system was described by means of several independent parameters, careful analysis would reveal non-ergodicity. That is, of all basically possible states only some are likely to occur in any single moment. Whether or not a given state is likely to arise, accordingly depends on the past or, more exactly, on the succession of states preceding the actual state – a phenomenon called path dependence. With regard to sustainability, path dependence plays a particularly important role in three respects. First, the wide variety of life forms in nature represents a large source of solutions for problems not only in the natural environment but also in the human sphere – for the assimilation of wastes, the production of food, and the design of pharmaceuticals, to mention just a few examples. Every species evidently represents a piece of knowledge that could potentially be useful for present or future generations. Against the backdrop of path dependence, however, it is also clear that the loss of any species leads to a loss of such knowledge that is irreversible. For every species is the outcome of a succession of phylogenetic stages in which the formation of every single stage is based on the existence of its respective predecessor – a fact that renders it impossible to reconstruct a species once it has been lost.

Second, even when knowledge is not directly acquired from models in nature, but derived through trial and error in the scientific process, this does not imply that all knowledge is equally accessible. Instead, technical knowledge generation is characterised by the formation of technological trajectories (Dosi 1982). Within such trajectories, knowledge acquisition occurs gradually – by the systematic small variation of single parameters and the selection of those variants showing the desired effect most markedly. Innovations proceeding along such a path are to some extent predictable but the marginal cost-to-effect ratio is subject to increase such that it becomes increasingly more difficult to make profitable innovations. An alternative route is the search for fundamental innovations leading to radical change between trajectories. While this approach has the potential for better profitability, it is characterised by a high degree of uncertainty representing a substantial burden for typically risk-averse people.

The third aspect of path dependence to be addressed here refers to the induced resistance-to-change and, thus, somehow relates to the second. It plays an important role in the discussion about technology development and is of central importance for the objective of this paper: the search of indicators for a sustainable technology development. Innovations and the introduction of new technologies often are the key instruments to the (temporary) avoidance or redressing of adverse environmental effects. However, even if negative external effects were completely internalised and the new technology turned out to be technologically and environmentally superior to the existing one, successful commercialisation and diffusion into the market cannot be taken for granted. A frequently quoted example for this kind of failure of a superior technology to prevail refers to the design of typewriter and computer keyboards (David 1985). Although the totality of users could benefit from the use of a better design that allows for a significantly higher writing speed, the traditional QWERTY keyboard is maintained because just for the first users of any new alternative, a deviation from the dominant design would cause costs that are much higher than the expected benefits. While network externalities are the relevant factor in the latter case, a variety of other effects will be identified in section 4 that lead to the lock-in of a conventional technology and, accordingly, to the lock-out of its superior challenger.

4 Indicators for second-order sustainability

In the preceding section, it was suggested that certain structural properties of a given technology can severely restrict the probability with which new innovations may become effective. The way in which these states of stability are sometimes discussed (David 1985) or modelled (Arthur 1988) in the literature could imply that such states of stability are omnipresent and, once they turn up, tend to persist for prolonged periods of time. Not surprisingly, some economists (e.g. Liebowitz and Margolis 1994) are convinced that positions like the preceding one crossly overstate the relevance of network externalities, as this would allow them to become the cause of almost ubiquitous market failure. In the latter debate, an intermediate position is taken by Witt (1997) who, while principally acknowledging the relevance of network effects, limits their general importance for the function of the market to certain restricted periods of time. So periods of stability tend to alternate with periods of instability where new networks can be formed. Such a period in which the direction of technological progress is flexible is referred to as a "window of opportunity" (Witt 1997). Disregarding these windows could severely hamper, if not completely inhibit, the introduction of any useful innovation. And even when, in the pursuit of sustainability, a new (sustainable) technology was successfully pushed by governmental regulation with no regard at the specific circumstances, the difference between stable and unstable phases would be worth a lot of money. It will therefore be the main objective of this section to identify all important factors and accordingly derive a set of indicators that allow political and other decision makers to make a well-founded judgement as to whether the preference for a potentially sustainable innovation is based on economic, social, and political feasibility.

The first set of factors will be economic ones. It will become evident in the following that the variety of relevant effects is wider and their respective time pattern more diverse than may have been implied by the repeated reference to network externalities in previous parts of this paper. Additionally, it is a special characteristic of many sustainable technologies that, beyond the competitive disadvantage frequently arising from their failure to internalise reduced external costs, the government typically plays a crucial role in overcoming existing barriers to competitiveness in the relevant markets. In doing so the government inevitably faces opposition from those whose interests are negatively affected: the incumbent industry and other groups paying the price for the measures taken. Typically, a government or policy makers in general are not inclined to neglect such an opposition unless the promoting forces from other parts of the society are sufficiently strong. More so, major techno-economic changes require a general openness or even a readiness to change (i.e. a phase of instability) on the part of the political system. For these reasons, the techno-economic factors will have to be supplemented by both, political and social factors. The selection of these criteria occurred on the basis of a priori theoretical plausibility considerations and ex post after the screening of relevant case studies (Sartorius and Zundel 2005). Due to the large number of relevant factors, it is not possible to present them here at length; for a more detailed discussion, the reader is therefore referred to Zundel et al. (2003, ch. 1).

4.1 Determinants of (in)stability in the techno-economic system

Economies of scale. Economies of scale are due to the fact that the benefit arising from employment of a more sophisticated machinery can more than outweigh its higher overhead cost if only the quantity of output can be increased sufficiently. They are typically measured on the firm level in terms of average unit cost as a function of output rate. While economies of scale a cause of strong competitive (cost) advantage, they are particularly relevant for new technologies which, at the beginning of their life cycle, cannot immediately engage into large-scale production.

Economies of scope. Economies of scope account for the realisation of synergies between different production lines. This includes among other things the common use of certain resources, intermediate products, or production facilities and, thus, requires a high degree of co-ordination. While economies of scope lead to important cost decreases for the established industry, the mutual dependencies between existing production lines make it even more difficult for a potential market entrant or a new technology to become competitive.

Learning by doing. Unlike the cases of economies of scale and economies of scope, the cost decreasing effect of growing experience in designing, constructing ('learning by doing'), and using production facilities ('learning by using') is a function of the cumulative output of a given branch of production over its entire history. The learning effects relevant in this context arise from incremental technical progress and are typically expressed as the percentage of cost/price reduction per doubling of the cumulative production output. While learning effects provide any new technology with a large potential for further cost reductions, they confront it with a high cost disadvantage in the beginning.

Sunk cost. Investment into a new technology can cause significant sunk costs if this investment renders useless an old technology in the same firm prior to its complete depreciation. Since sunk costs represent opportunity costs of the new technology, they cause a systematic disadvantage for any new technology. While the latter argument does not come to bear in competitive markets, it is indeed relevant whenever market access is restricted by other causes. The rate of capitalisation in the relevant industry and data about the investment cycle can be used to assess sunk costs; however, the analysis needs to be supplemented by the competitive structure of the industry in question (see below).

Network externalities. Network externalities refer to the fact that the utility derived from the use of a given technology is positively correlated with the number of its users. Alternatively, a technology can be subject to network externalities if, rather than constituting a network itself, it relies on another technology that forms the network in its turn. Whether or not network externalities actually constitute an entry barrier for a new technology, depends on the dependence of the latter on an existing (technology) network and, if so, on their mutual compatibility. The weaker the dependence and the better the compatibility, the smaller the competitive advantage that can be drawn from network externalities by each competitor.

Market structure. In many markets, the number of market entries is limited by specific (declining average) cost structures or by governmental regulations, giving rise to natural or regulated oligopolies or monopolies. Although this does not exclude competition in principle, such market structures will provide the corresponding firms with strong incentives to maintain the existing market barriers, to engage into strategic interaction with other market participants for the realisation of monopoly rents, and to neglect innovative activities. Therefore, any non-competitive market structures will strongly stabilise the existing technology at the expense of potential competitors.

Potential versus risk. Marginal returns within any given technological paradigm tend to decrease in time. In order to replenish their earned innovation rent and, thus, maintain their current profit margins within a competitive market environment, entrepreneurs therefore have to complement their technological portfolios occasionally with more radical innovations. Since more radical innovations are associated with higher risk, an (expected) strong potential (including its regulatory conditions) will be decisive for the success or failure of a new technology to be adopted.

Demand. In order to be considered an economic substitute for an existing technology, a new technology will have to fulfil certain functions of the former that are crucial for attracting the attention and raising the specific demand of those consumers and investors that would otherwise buy the established technology. But this by no means implies that both technologies have to resemble each other in most or even all of their remaining properties. Since, after comparing two almost equivalent technologies, most people would probably buy the established version they are more familiar with, a new technology therefore has to fulfil as many extra-functions as possible to overcome this inertia.

Niche markets. If the entry barrier for a new technology is high, it may need a long period of subsidisation until general competitiveness is achieved. At the same time, partial competitiveness may be achieved under certain, for instance geographically or culturally specified conditions. Such an environment in which the new technology is economically viable despite its marked competitive disadvantages in the general market is called a niche market. The existence and the extent of niche markets can be decisive for reaching competitiveness of a new technology in general. In the same vein, artificial creation of such a niche market through governmental regulation can be an important approach to the successful implementation of a new technology.

4.2 Determinants of in-/stability in the political system

The basic characteristics of the political system generally play an important role in allowing a new, more sustainable technology to prevail. As a precondition for this to happen, the political system either must be in favour of the new technology from the beginning or it needs to be destabilised itself in the first place. While in the former case, structural characteristics of the political system play the most important role, both structural and procedural aspects are important in the latter. The following enumeration will begin with the structural factors and then shift to the procedural ones.

Institutional embeddedness. Many technologies, particularly those related to environmental protection, are subject to substantial political regulation that determines which external effects a technology is allowed to exert and which (and how) others must be avoided. In this context, the design of, and the mutual interaction between, the relevant institutions can greatly influence the competitive position of an innovation as opposed to the established technology. If, for instance, the regulatory restrictions specifically refer to an existing technology as the state of the art in solving an environmental problem, this technology is strongly stabilised as opposed to all innovations that approach the problem in a different way and, thus, have to pass approval and licensing procedures in order to conform with the regulation.

Interest groups. While it is a matter of political culture how influential corporate bodies or individual actors can be in principle (see e.g. the dependence of the government or the political administration on any kind of support from certain industries), it depends on the specific circumstances which effects they actually give rise to. Basically, the power of an interest group is known to be crucially dependent on the size of the group, the homogeneity of its interests, its organisation, and the resources it controls (Olson 1965). Other important factors are the economic relevance of the industry or its history and its cultural integration. Particularly in mature industries with strong market power, lobbying may even pay for single firms as from their perspective, investing in a useful regulatory institutional environment may be more profitable than investments in technological innovations (Berg 1995). As a consequence, most lobbying activities will tend to stabilise the established technology.

Asymmetry of knowledge. For the solution of environmental problems, governments and political administrations need external advice. So long as the problem has not attracted too much public attention, it is most convenient for the political administration to try to obtain the necessary information from the industry that caused the problem. According to the life cycle theory of bureaucracies, initially independent (regulatory) authorities will then successively merge their interests with those of the established industry (including the technological trajectory it represents) (Martimort 1999). This "regulatory capture of bureaucracies" often leads to quick and at most half-hearted solutions predominantly related to the dominant technology. By contrast, more radical changes can only be expected, if the necessary knowledge comes from more independent sources – notably state-financed scientific research.

Parliamentary majorities. Especially more radical changes are often not unanimously supported since the improvement of the situation of some people goes at the expense of others. Even if its basic attitude would tend to render a government or a political party supportive of this change, its actual realisation will ultimately depend on the strength and stability of the majority on which the politically acting group can rely. From this perspective, a large, stable majority basically opens the potential for more radical changes than does a minute or unstable one.

Election cycle. One of the most prominent stylised facts in political science states that more radical political changes usually occur at the beginning of an election period while incremental changes, if not political standstill, follow at the end (Troja 1998). With regard to environmental innovations this implies a potential for greater instability of the established technology (i.e. a political window of opportunity) in the post-election period. Unfortunately, empirical tests so far failed to confirm this effect of the election cycle (Horbach 1992). A special popularity of environmental regulation or an eminent problem pressure could be reasons for this. In

Germany, the temporal alternation between state and federal elections additionally renders the distinction between pre- and post-election periods obsolete. Finally, it has to be recognised that many aspects of environmental innovation policy are consensually negotiated and, therefore, unsuitable as topics for an electoral campaign.

Singular constraints. The costs and, thus, the scope of each regulatory measure is subject to a budget constraint. However, the latter is itself the result of negotiations between a variety of parties, each wishing to appropriate the largest share of the budget at disposal. While in many cases, the power of the interest groups behind technologies influences the allocation of governmental resources, this is not a natural outcome. In the end, it may depend on the social appreciation of environmental protection or the reputation of the involved parties whether the incumbent industry can defend its subsidies or has to share it with its more sustainable competitors. In this respect, a sudden change could also be brought about by singular (i.e. exogenous) events like political scandals and environmental or other catastrophes.

Decision-making procedures. Since it is not possible here to extensively analyse the entire political decision-making process, just a few criteria will be presented that may allow for a basic characterisation of the procedural aspects of a political system with regard to the stabilisation or destabilisation of a specific technology.

1. It is an important aspect of political culture whether the initiatives for regulatory acts typically come from single actors (e.g. president, members of parliament) or major bodies (government, parties, or the parliament). Individual-based initiatives tend to give rise to more radical (i.e. destabilising) changes than those of (more consensus-oriented) corporate bodies.
2. The relation between the legislative bodies and the executive administration determines whether a regulation is generally enacted by means of a law that has to pass a lengthy parliamentary approval procedure or whether this can be done by referring to an ordinance that is quickly adopted by the political administration.
3. Obligatory reassessment and the enactment of resubmission cycles ensure that the existing regulation does not lead to the stabilisation of the respectively benefiting technology.
4. Another important aspect of the political culture refers to the existence and influence of corporate structures (e.g. industry associations and labour unions); they typically refer to, and stabilise, established technologies.

5. Participation of larger parts of the society (e.g. NGOs, public research institutes) in the search for more sustainable solutions will not only facilitate the search for knowledge but also increase and widen the support for (often more radical) solutions.
6. Finally, it is important how a country is incorporated into supranational structures (e.g. EU, WTO). While this limits a country's possibility to implement innovations in an idiosyncratic manner, it broadens the scope and efficacy of many sustainable innovations.

4.3 Factors of change in the socio-cultural system

Public attention to a (perceived) problem and subsequent worry about its potential consequences play a key role in provoking political reactions directed to solving the problem or, at least, alleviating its consequences. This is all the more true in the context of environmental protection since due to their long-term relevance and public-good nature, environmental problems and their solutions are rarely issues that allow a politician to derive major benefits for himself. While awareness and concern by a considerable part of the population is neither sufficient nor necessary for political action to be initiated, their lack will usually lead to a failure or, at least, major delay in acting accordingly.

Mass media play an important role not only as transmitters for the corresponding information but also for the assignment of meaning and valuation to the underlying problem. The relation between the media and their readers, listeners, or watchers is characterised by mutual interaction giving rise to positive and negative reinforcement The scientific verification of an environmental problem which often stays at the beginning of such an 'issue attention cycle' (Downs 1972), is identified through scanning the scientific literature for relevant keywords and trying to identify seminal publications through the tracing back of references. On the other hand, public concern about these problems can be measured to some extent by counting relevant articles in newspapers and reports in other mass media. Additionally, it may be necessary to account for the more qualitative aspects of concern and valuation, as the authors of relevant articles often differ in their basic attitude towards a given environmental problem. It is also important to realise that the attention of mass media to any given problem usually tends to decline more rapidly than the attention of the public in general.

Table 1. Factors determining the stability or instability in each of the three subsystems and the indicators used for their operationalisation

Effect	Indicators	Operationalisation
Techno-economic system		
Economies of scale		cost (or price) development as a function of actual output
Sunk costs	average capitalisation of the industry	statistical data
	identification of investment cycles	recurrent phase-shifted cycling of prices and investment
	political regulation	cost of retro-fitting after regulation, delayed investment due to expectation of uncertain measures
Economies of scope	pattern of interactions between production lines	number and relevance of interactions between the old (new) technology and the entire production network
Learning by doing		cost (or price) development as a function of cumulative output
Network externalities	direct competition with (an)other network(s)	market share(s) of the competitor(s), availability of gateway technologies
	need for compatibility with complementing infrastructure or periphery:	
	• existence of public standards	which requirements are met?
	• availability of an adapter	cost of the adapter, legal admission possible, payable royalties
Market structure	degree of competition as a function of market concentration	market share of the biggest firm(s), Herfindahl-index, legal regulations
Potential vs. risk	riskiness ↔ availability of capital	marginal interest rate, capital share of venture capitalists
	problem solving capacity ↔ realisation of an innovation rent	technical properties, associated costs
Extra-demand	readiness to pay for extra-functions	market research
	existence of natural niche markets	higher prices, non-applicability of the established technology
	creation of artificial niche markets by means of regulation	(eco-)taxes, tradable certificates, cost of retro-fitting the old technology

Table 1. (cont.)

Effect	**Indicators**	**Operationalisation**
Political system		
Institutional embeddedness	Subsidies	financial support, tax breaks
	Protection	duties, other barriers to trade
	norms and standards	specificity of specification
Interest groups	resources under control (power)	number and economic importance of represented firms/sector
	structure of the basis; degree of homogeneity	market shares, concentration index
	influence; earlier success	(qualitative)
Asymmetry of knowledge	influence of (incumbent) industry in hearings	(qualitative)
	number of industry-independent research institutions/projects	number, financial support, number and size of commissioned projects
Parliamentary majorities	stability of majorities	size of majority, stability of constituting coalition (number and relation of parties)
Election cycle	distance to the next election	ditto
Singular constraints	political scandals Catastrophes	deception by possible interest holders accidents, unexpected discoveries
Decision-making procedures	probability of legislative initiatives	number and relevance of potential initiators, number of actual cases
	legislative vs. administrative regulation	number of laws referring to ordinances, actual number of ordinances
	reassessment and resubmission cycles	deadlines, frequency, possible consequences
	corporate structure	number, size, and frequency of political involvement of corporate organisations
	Participation	frequency and extent of incorporation of political "outsiders" (e.g. NGOs) into the decision process
	supranational structures	share of regulation that is not subject to national legislation

Table 1. (cont.)

Effect	Indicators	Operationalisation
Socio-cultural system		
Scientific verification of threat to sustainability	relevant publications in scientific literature, contributions to conferences	number of relevant articles (keyword search) in journals or conference proceedings and monographs; identification of seminal articles and quotation circles
Public concern about lack of sustainability	relevant articles in newspapers, reports in broadcast	number of articles/reports over time
Public acceptance of possible solutions	formation of major protest campaigns	number and size of campaigns

4.4 Integration of the indicators

After elaboration of a large, comprehensive set of indicators in the preceding parts of this section the question naturally arises as to how an integration of these indicators can be achieved. The first restriction to the achievement of this goal comes from the fact that most but not all indicators can be assessed in quantitative form. To determine their effect on the stability or instability of the established technological regime, it is necessary to compare the latter with its more sustainable alternative and to figure out the meaning of this difference. Here, a small difference in terms one property can be more important than a large difference in terms of another. So, representation of the entire comparison by a single pair of numbers is impossible.

The latter problem also applies to all those indicators that are indeed available as single figures. Even if these figures are expressed in the same dimension (e.g. monetary value), their meaning for the ultimate goal is quite different (compare sunk costs and size of niche market). As a consequence, any comparison can in the end only be of qualitative nature.

The next problem refers to the aggregation of the different factors. In the techno-economic sphere, all factors essentially work in parallel. High sunk costs add to the stability of the incumbent technology as well as does extended learning. Niche markets for the new technology on the other hand destabilise the incumbent. None of these factors relies on another one to become effective. So, even if one effect became zero, the other factors would remain unaffected. This mode of aggregation is called additive.

By contrast, in the socio-cultural system, (scientific) verification of an environmental problem is a necessary (but not sufficient) prerequisite for the formation of public concern. So, without discovering the problem, there will not be any concern. Conversely, public concern alone sometimes is little effective until the exact causes for an environmental problem are scientifically verified. So, both factors work in sequence with the combined effect yielded by multiplying the single constituents.

In the political system, both effects are found. While structural and procedural factors in general appear to complement each other in a multiplicative way, the specific structural (or procedural) factors tend to work in parallel.

With regard to the relationship between the entire systems, the political system not surprisingly is of central importance because in the end, it brings about the regulation. However, the political system hardly works on its own; it needs impulses from the other systems: destabilising impulses (for the existing technological regime) come from the society disapproving the lack of sustainability and/or from the new, more sustainable technological or institutional alternatives; opposite stabilising impulses come from the incumbent industry that caused the environmental problem and the loss of sustainability in the first place. Figure 1 summarises how the composite indicator of sustainable technology development is constructed from its constituents.

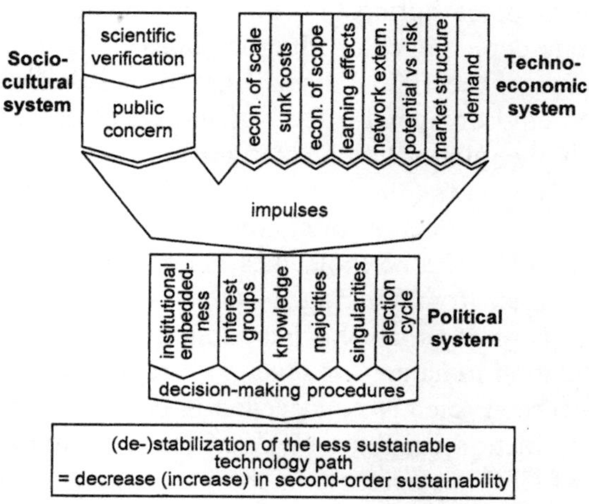

Fig. 1. Reconstruction of a comprehensive indicator for the successful implementation of sustainable innovations from its constituent factors in the techno-economic, political, and social sphere

5 Application of the new indicators: the phase-out of CFC

The indicator of sustainability or, more precisely, sustainability-directed technology development that has been developed in the preceding section, significantly differs from other indicators in referring not so much to the environmentally relevant properties of specific technologies, but to the entirety of the system properties that allow those technologies to become effective by entering the market in the first place. In order to use this indicator strategically, it would be necessary to first check all its components for relevance in a given context of an unsustainable technology and its potential substitute(s). Then all significant aspects would have to be assessed in terms of stabilising or destabilising effects and their changes in time. Finally, after specifying the mode of interaction between the relevant components, aggregation would yield a kind of time profile of in/stability reflecting the ease of transition from an established to a new technological path. Since at least some of the components are subject to influences by the political system, the whole analysis provides useful hints to the design of a policy that reaches sustainability targets most effectively.

Though under way, such ex ante studies are not yet completed. So, in order to illustrate the operability of the proposed method of analysis, I will refer to the ex post analysis of a rather successfully regulated technological transition that took place during the last quarter of the 20th century: the phase-out of ozone-depleting chlorofluorocarbons (CFCs)[2].

In the beginning of the 1970s, a small group of scientists became concerned about the environmental effects of the emission of chlorine compounds into the higher atmosphere. Among the major impacts of this group was Molina and Rowland's (1974) detection of a chemical mechanism potentially leading to the depletion of stratospheric ozone by chlorine atoms originating from CFCs. Ozone molecules were known to be essential for blocking UV radiation from entering those parts of the atmosphere where they would cause harm to organisms including humans. Although these results lacked validation in nature for several years, this community of environmentally concerned scientists succeeded in conveying their findings to environmental protection groups which reacted by initiating a campaign against the use of CFCs as aerosols in spray cans. Many consumers complied by not buying spray cans before a law prohibiting this usage of CFC was enacted in 1978. This led to a temporary reduction in CFC emissions

---

2 For a more comprehensive analysis of this case and for other case studies refer to Sartorius and Zundel (2005).

which was soon compensated by the increasing use of CFC in uses other than as aerosols (Meadows et al. 1992).

Enacting a law against CFC in spray cans did not require too much pressure after it turned out that the substitution for CFC even led to cost savings. Another reason for this success was the particular reliance of U.S. politicians on scientific arguments. But in order for the U.S. government to take more extended measures to reduce CFC emissions, the evidence in nature for the Molina-Rowland hypothesis was simply too weak and the opposition against such measures was too strong. As a case in point, a DuPont executive testified before Congress in 1974 that the "chlorine-ozone hypothesis is at this time purely speculative with no concrete evidence to support it." However, "[i]f creditable scientific data ... showed that any chlorofluoro-carbons cannot be used without a threat to health, DuPont will stop production of these compounds." (Meadows et al. 1992). At the same time, it was quite clear that the USA would be the major stake-holder in all measures concerning CFC since they were both the biggest producers and the biggest consumers of CFC. Due to significant differences even within the Reagan administration, however, it was not clear until the second half of the 1980s what position (positive or negative) would eventually be adopted in this respect. It was the U.S. Environmental Protection Agency (EPA) and the U.S. state department's bureau of Oceans and International Environmental and Scientific affairs (OES) that tended to adopt the critical scientists' position. On the other hand, interest groups and governmental offices related to chemical industry tended to adopt the viewpoint that important and far-reaching governmental regulation in the field could not be justified, not to mention the Reagan administration's general attitude was against any kind of regulatory intervention.

Then, in 1984, the first evidence for a big 'ozone hole' over Antarctica was found. Scientists of the British Antarctic Survey measured a 40 percent decrease in ozone in the stratosphere over Antarctica. While it took until later in 1987 that the causal relation between CFC emission and the ozone hole was finally established, the existence of the ozone hole was sufficient to initiate a powerful movement that eventually led to the ban of CFCs. Internationally, an important role in the latter process was played by the UN Environment Programme (UNEP) which organised a series of big international conferences intended to make a rigorous assessment of the remaining uncertainties of, and provide solutions to, the relationship between ozone depletion and CFC emission. When, as a result, evidence of the ozone-depleting effect of CFC had finally become strong enough to serve as an argument in favour of CFC regulation, especially two events led to a successful agreement in 1987. First, DuPont honoured the pledge it had made more than a decade ago and came to share the critical scientists' concern about CFC-caused ozone depletion. This "change in mind" of the biggest producer of CFC in the U.S. and world-wide led to a collapse of the U.S. industrial opposition against CFC regulation. Second, as a consequence of the discovery of the ozone hole and of other negative ecological impacts (e.g. the accidents in Schweizerhalle and in Chernobyl), green parties particularly in Germany became more influential. Together with a change in the presidency of the European Commission, this gave rise to a turn in the EU attitude that originally opposed CFC regulation.

In the end, international agreement on the Montreal protocol led to a two-step reduction of CFC production of 20% by the year 1993 of a total of 50% by 1998. Three years later (1990) in London, an amendment was ratified by 92 countries yielding a complete phase-out by the year 2000 and another two years later (in the Copenhagen amendment) the phase-out was advanced to 1996. This total ban of these chemicals within a single decade is all the more surprising in view of the economic relevance of CFCs (the USA alone produced almost one million tons of CFC each year).

With regard to the analysis in terms of stability and instability, the CFC story can be divided into two parts terminating in the ban of CFC-containing spray cans (in 1978) and in the Montreal protocol (in 1987) and its successors, respectively. In each part, the political system played a central role in the ban of CFC since without the basic readiness of the political system (and the corresponding window of opportunity being open), regulation would not have taken place. However, in both cases, additional support from the social system (i.e. an open window there) was useful, if not essential, in several respects. First, the scientific community played a crucial role in the social system by discovering the environmental problem associated with CFC emission and directing people's awareness and concern to it (Grundmann 1999). Second, a strong impulse pro regulation from the social system was necessary (though not always sufficient) to counterbalance contra regulation impulses coming from the economic system. This effect was even enhanced by the demonstration of a significant proportion of society that the environmentally harmful goods or services are indeed unwelcome. Third, the open window in the social system served as a legitimisation and incentive for policy makers to pursue regulatory measures against opposing forces from within the political system. Altogether, the social window of opportunity the opening of which was caused by the discovery and confirmation of the ozone-depleting effect of CFC, in its turn gave rise to an opening of the political window in the first place.

The following factors were crucial for the readiness to change of the political system. While the majority for the Democrats had been responsible for the enactment of the Clean Air Act and the ban of CFC in spray cans in the first phase, it was the initiative of individuals like U.S. chief negotiator Richard Benedick and a scandal in the EPA that led to the reconstitution of a pro-regulation regime despite the Republican government after 1983. Since the interest group contra regulation consisted only of a few chemical manufacturers with DuPont representing the biggest player, they were powerful enough to prevent major regulation before 1987; however, the alliance immediately collapsed after DuPont changed its attitude in 1986. Finally, the increasing role of environmental policy in some European countries and special ambitions of the former German chancellor Kohl in the EU led to a change in the supra-national actor constellation that allowed for the agreement in the Montreal protocol. By contrast, other factors like institutional embeddedness or knowledge asymmetries did not exert a significant effect.

Whether or not, at last, the economic window of opportunity was open for a regulation crucially depended on the cost-benefit calculus employed. Here it is important to distinguish between effects on the level of the economy which were more directly relevant for the response of the political system and effects on the firm or industry level that were crucial in terms of the pressure exerted on the political system. In the latter case, the most important costs of a regulation of CFCs were the sunk costs associated with the then obsolete production facilities (for CFCs) and the risk associated with the introduction of substitutes whereas the decisive benefit resulted from avoiding potential liability suits of those people that would eventually turn out to suffer from CFC-related skin cancer. Other techno-economic factors like economies of scale, economies of scope, learning, and network effects did not play such a crucial role as substitutes for CFCs were readily available with regard to production as well as demand[3]. Eventually, it was the confirmation of the direct link between the emission of CFC on the one hand and the break-down of the ozone layer and the concomitant increase in the irradiation of the earth's surface with ultraviolet light on the other, that forced the CFC producers to give up their opposition. While a variety of different substitute technologies was engaged in competition with CFC, it could be shown that those CFC substitutes supplied by the chemical industry initially benefited from first-mover advantages, that is, from the fact that they were in place first. Due to the regulation method employed (i.e. CFC emission trading), however, this advantages did not give rise to the displacement of other substitutes. Thus,

---

[3] It was certainly in support of the phase-out process that substitution took place in several steps with HCFCs first replacing CFCs, then HFCs replacing the latter two, and finally, at least in some applications, hydro carbons, CO2 or ammonia replacing HFCs.

the techno-economic window could be kept open to increase the number of alternatives among which selection was supposed to take place.

In the end, the successive destabilisation of the CFC regime (and opening of the corresponding windows of opportunity) in each, the social, techno-economic, and political system has led to one of the most prominent cases of successful innovation policy towards sustainability.

6 Conclusion

In particular radical innovations can be important means to the achievement of improved sustainability. Due to the existence of path dependencies, however, the transition from one technological trajectory to another, more sustainable one is often impeded by significant barriers. Fortunately, these barriers are by their nature subject to substantial changes; so, it makes sense to carefully distinguish between periods of stability (with high barriers) in which the given trajectory can hardly be left and periods of instability (characterised by low barriers) where a new trajectory can be reached more easily. With respect to sustainability, the latter distinction is particularly important for two reasons. First, more sustainable innovations often rely on governmental regulation. In periods of instability, the economic burden arising from this regulation will be much lower than in periods of stability; so, a given budget will yield a much better sustainability effect in the former case than in the latter. Second, due to the complexity and changes in their respective environments, innovations are generally associated with fundamental uncertainty such that it becomes impossible to predict the degree of sustainability resulting from specific innovations in the long run. Under these circumstances, it is essential to allow for rapid change with the possibility to select between a variety of different trajectories within a process of trial and error. Sustainability as viewed from this evolutionary perspective may therefore better be understood as the general capability to readily change between different technological trajectories.

In order to undergo successful diffusion, most sustainable innovations rely on regulatory measures especially in the beginning of their (economic) lifecycles. When looking for the factors determining periods of (in-)stability, the political system enacting this regulation therefore is of central interest. However, while basically allowing for the convergence of both technological progress and sustainability, the political system itself can neither give rise to the search for sustainability nor bring about the appropriate innovations in the first place. This is where the socio-cultural and, of course, the techno-economic sphere itself enter the focus of attention as emitters of positive impulses. Additionally, negative impulses like those coming from the incumbent industry need to be taken into account. After all, a series of factors (and corresponding indicators) could be identified which after proper weighting and prioritisation allow to make an estimation whether, and possibly when, the incumbent industry is sufficiently destabilised and the political system rendered sufficiently favourable to the new, more sustainable technology such that a transition to the preferred trajectory is possible without too much effort.

7 References

Arthur WB (1988) Self-Reinforcing Mechanisms in Economics. In: Anderson PW, Arrow KJ, Pines D (eds) The Economy as an Evolving Complex System. Redwood City

Berg C (1995) Technologischer Fortschritt und ökonomische Regulierung. Frankfurt/Main

Daly HE (1990) Toward some operational principles of sustainable development. Ecological Economics 2: 1-6

David PA (1985) Clio and the economics of QWERTY. American Economic Review, Papers and Proceedings 75(2): 332-337

Dosi G (1982) Technological paradigms and technological trajectories. Research Policy 11: 147-162

Downs A (1972) Up and Down with Ecology. The Issue-Attention Cycle. The Public Interest 28: 38-50

Erdmann G (2000) Innovation, Time, and Sustainability. (Paper presented at the International Conference "Towards Environmental Innovation Systems". Garmisch-Partenkirchen)

Grundmann R (1999) Transnationale Umweltpolitik zum Schutz der Ozonschicht, Frankfurt. New York

Hartwick J (1978) Substitution among exhaustible resources and intergenerational equity. Review of Economic Studies 45: 347-354

Hayek FA (1978) Competition as a Discovery Procedure. In: Hayek FA (ed) New Studies in Philosophy, Politics, Economics, and the History of Ideas. London

Horbach J (1992) Neue Politische Ökonomie und Umweltpolitik. Frankfurt/Main, New York

Jaenicke M (1999) Policy patterns and innovation impacts: an international comparison. (Paper presented at the International Conference 'Innovation-oriented Environmental Regulation – Theoretical Approaches and Empirical Analysis'. Potsdam)

Klemmer P, Lehr U, Löbbe K (1999) Environmental innovation. (volume 3). Berlin

Liebowitz SJ, Margolis SE (1994) Network Externality: An Uncommon Tragedy. Journal of Economic Perspectives 8(2): 133-150

Martimort D (1999) The life cycle of regulatory agencies: dynamic capture and transaction costs. Review of Economic Studies 66: 929-947.
Meadows DH, Meadows DL, Randers J (1992) Beyond the Limits. London
Molina M, Rowland S (1974) Stratospheric Sink for Chlorofluoromethanes: Chlorine Atom Catalyses Destruction of Ozone. Nature 249: 810-812
Nelson RR, Winter SG (1982) An Evolutionary Theory of Economic Change. Cambridge (MA)
Norgaard R (1994) Development betrayed. London
OECD (1997) OECD Proposed Guidelines for Collecting and Interpreting Technological Innovation Data – Oslo Manual. Paris
Olson M (1965) The Logic of Collective Action. Cambridge (MA)
Opschoor JB, Reijnders L (1991) Towards sustainable development indicators. In: Kuik O, Verbruggen H (eds) In Search of Indicators of Sustainable Development. Dordrecht
Pearce DW, Atkinson GD (1993) Capital theory and the measurement of sustainable development: an indicator of 'weak' sustainability. Ecological Economics 8: 103-108
Pearce D, Markandya A, Barbier EB (1989) Blueprint for a Green Economy, London
Rennings K, Wiggering H (1997) Steps towards indicators of sustainable development: Linking economic and ecological concepts. Ecological economics 20: 25-36
Rennings K (2000) Redefining innovation – eco-innovation research and the contribution from ecological economics. Ecological Economics 32: 319-332
Sartorius C (2002) The Relevance of the Group for the Evolution of Social Norms and Values. Constitutional Political Economy 13: 149-172
Sartorius C (2003). An Evolutionary Approach to Social Welfare. London
Sartorius C, Zundel S (eds) (2005) Time strategies, innovation and environmental policy. Cheltenham (forthcoming)
Schumpeter JA (1934) The theory of economic development: An inquiry into profits, capital, credit, interest, and the business cycles. Cambridge (MA)
Solow RM (1986) On the intergenerational allocation of natural resources. Scandinavian Journal of Economics 88(1): 141-149
Troja M (1998) Umweltpolitik und moderne Ökonomik: Der Beitrag der Neuen Politischen Ökonomie und der Neuen Institutionenökonomik zur Erklärung umweltpolitischer Entscheidungsprozesse. Studien zur internationalen Umweltpolitik. Münster
Witt U (1997) 'Lock-in' vs. 'critical masses' – industrial change under network externalities. International Journal of Industrial Organization 15: 753-773
Zundel S, Erdmann G, Kemp R, Nill J, Sartorius C, Weiner D (2003) Zeitstrategien ökologischer Innovationspolitik – der Forschungsansatz. In: Horbach J, Huber J, Schulz T (ed.) Nachhaltigkeit und Innovation. Rahmenbedingungen für Innovationen zum nachhaltigen Wirtschaften. München

Indicators for Lead Markets of Environmental Innovations

Marian Beise and Klaus Rennings

1 Introduction

The response of consumers to new products is a crucial factor for their success. And the success of new products, creating new markets, are of paramount importance for innovation. It is expected that the market's impact on innovation will grow in the future, and the majority of managers expect that markets will become more receptive for introducing new products (ITT 2003). In this context innovation policy needs a deeper understanding why innovations are adopted by pioneer countries and diffuse from country to country. These processes are the issue of the "lead markets" concepts. It explains competition between different innovation designs, early adoption in lead markets and the following global diffusion (Beise 2001).

In this paper we extend the lead markets approach to environmental innovations, emphasising the important role of regulation for these type of innovations. In the context of the indicator discussion of this book, we focus on international diffusion curves of environmental technologies as indicators for lead markets and successful innovations.

National markets vary in their flexibility concerning the adoption of a given innovation. Lead markets are not necessarily the countries that developed a new technology. Others may adopt it first due to specific conditions. The price and cost structure of a national market can be encouraging for certain types of innovation. For example, automation technologies develop faster in countries with relatively high labour costs, and energy saving innovations in countries with higher energy prices. Concerning environmental innovations, these price and cost structures largely depend on regulation.

The paper is structured as follows: In section 2 we present a general model of lead markets developed by Beise (2001). In section 3 we extend this model to eco-innovations, considering their peculiarities, and develop a framework for further analyses. Due to the peculiarities of eco-innovations we identify environmental regulations and policy diffusion as key lead market factors. In section 4, two case-studies are analysed with the derived framework: the emergence and international diffusion of wind energy generation and fuel efficient technologies for passenger cars. Section 5 draws some conclusions.

2 The general lead markets approach

2.1 The international diffusion of innovations

A first step in analysing the international success of eco-innovations and the respective policies is to study the determinants of the international diffusion of non-environmental innovations. Looking at the diffusion of globally successful innovations, one can observe, that many innovation designs became internationally successful after they have been preferred and adopted by a single country. The facsimile machine, for instance, was adopted in Japan before it became the globally preferred design for text-based telecommunication. Cellular phones were widely adopted in the Nordic countries first and the US led the adoption of the personal computer. The leading user country normally stays ahead in the penetration rate for a long time, supplying the firms with long-term user feedback and market knowledge, enabling them to constantly improve the innovation and remain in the lead. Figure 1 exhibits the typical international diffusion pattern of a specific innovation design. Countries that are first in adopting an internationally successful innovation can be called lead markets, the following countries the lag markets.

Differences between lead and lag markets cannot simply be answered with reference to a lesser degree of "innovativeness" in the lagging countries. While export success of a country's firms has previously been explained mostly by leads in technological knowledge, demand and market conditions that lead to an early adoption of innovations are vital factors for the international competitiveness of countries as well (see Porter 1990). Historic studies of globally successful innovations such as Vernon (1966), Franko (1976), Tilton (1971) and several of the case studies collected in Mowery and Nelson (1999) have indeed found that the origin of the international competitiveness of a country is a demand gap and that this demand gap has caused the technology gap observed after the product became established world-wide. The technological gap is based mainly on experience in production (learning-by-doing) and usage (learning-by-using). In contrast, discoveries and inventions often occurred in countries other than the country where the innovation was first widely adopted. There, firms could usually use technical knowledge from abroad to match local demand.

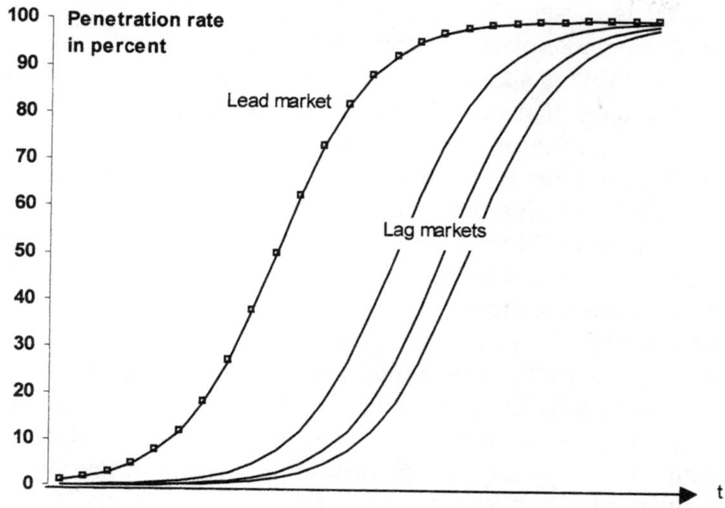

Fig. 1. The international diffusion pattern of an innovation design

Lead markets are countries that adopt successful innovations quickly, despite the fact that they did not necessarily invent the technology. Frequently, users in other countries have adopted rival innovation designs before which never became a success abroad. But only the innovation design adopted in the lead market becomes adopted in other countries and finally supersedes designs previously preferred in other countries. For instance, the telex system was adopted before the market breakthrough of the facsimile machine hit the road; in France the online-service "Minitel" similar to the Internet was adopted in the early 1980s before the Internet took off in the United States. This illustrates, that innovations that have been successful in their home countries have to compete on the world market against other innovation designs that are favoured by other countries due to their specific environment or market conditions.

The introduction of competing innovation designs is a factor for understanding why the early adoption of an innovation in one country sometimes leads to an export success and sometimes not. An innovation design is a specification or configuration of an innovation idea. Different designs of an innovation fulfil the same function but can have different modes or specifications (see Utterback 1994). For instance, an IBM, an Apple Macintosh or a Sinclair computer are different designs of a personal computer. The GSM cellular telephone, personal satellite telephony and pagers are different designs of mobile communications. A dominant design is defined as a design that is adopted by a majority of users, a design that wins the "allegiance of the marketplace" (Utterback 1994). A globally dominant design is the design that is adopted by most countries, in contrast to nationally dominant designs, that are only widely adopted within one country.

Lead markets are countries that first adopt a globally dominant innovation design; they lead the international diffusion of an innovation and set the global standard. For instance, the mass market for cellular mobile telephony emerged in the Nordic countries before a joint-European cellular system became the world standard in mobile telephony. And parallel with the United States leading the usage of PCs the IBM-Microsoft-Intel specification prevailed on the world market as the global dominant design of personal computers. On the other hand, several innovation designs became widely adopted in one country but never became an export success and even squeezed out of their home market years later by a global dominant design. Countries that are early adopters of an innovation design that is not adopted by other countries can be called idiosyncratic markets. A firm responding to this idiosyncratic demand can achieve temporary innovation success in these local markets but later has to switch to the dominant design, thus losing its pioneer advantages. Figure 2 includes the penetration rates of a competing innovation design that was initially adopted by a lag country that switched to the lead market design later. This pattern shows that lead markets are not necessarily the most innovative markets.

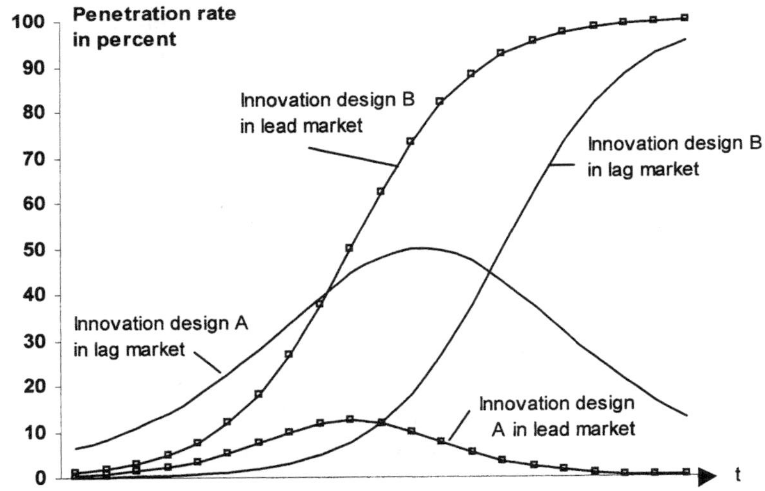

Fig. 2. The international diffusion pattern of competing innovation designs

2.2 Lead market factors

Beise (2001) has identified a typology of five basic groups of advantages of a lead market. The five lead market factors are as follows:

- the price advantage,
- the demand advantage,
- the transfer advantage,
- the export advantage and
- the market structure advantage.

A price advantage arises from national conditions that result either in relative price decreases of a nationally preferred innovation design compared to designs preferred in other countries or in anticipation of international factor price changes. Countries can gain a price advantage if the relative price of the nationally preferred innovation design decreases so that differences in demand preference to foreign countries can be compensated. This price mechanism is the centrepiece of Levitt's (1983) globalisation hypothesis, according to which the consumers in foreign markets "capitulate" to the attraction of lower prices and abandon their initial endowments of goods. Price reductions are mainly due to cost reductions based on static and dynamic economies of scale. Nation-specific factors of economies of scale are market size and market growth. Another price advantage emerges from anticipatory factor prices: the lead market demands innovations that are induced by factor price changes which later occur globally.

Demand advantages are caused by national conditions resulting in the anticipation of the benefits of an innovation design emerging at a global level. A good example are off-grid solutions in the energy and telecommunication sector. Such innovations are more beneficial and thus more likely to be adopted first in industrialised countries with large landscapes and a low population density, as e.g. Scandinavia. When other countries catch up, they demand the same innovation that was already used in the country at the forefront of the trend. Another example are trends related to environmental problems such as climate change. Some countries suffer more from the risks of rising temperatures (e.g. countries with above average risks of flooding like the Netherlands) than others and will thus anticipate this trends earlier.

Transfer advantages are national conditions that increase the perceived benefit of a nationally preferred innovation design for users in other countries or by which national demand conditions are actively transferred abroad. The perceived benefit increases when information on the usability of the innovation design is made available. The first adoption of an innovation of unknown merit reduces the uncertainty and therefore the risk for subsequent adopters initiating a bandwagon effect. This is also called the demonstration effect of adoption (Mansfield 1968). A country can have a transfer advantage if its market context supports increases in the perceived benefit of a nationally preferred innovation design for users in foreign countries. Diffusion theory suggests that the international diffusion of durable goods depends on the intensity of communication between two countries (Takada and Jain 1991). The lead market could therefore be the country that has the strongest communication ties with other countries.

Conditions that support the inclusion of foreign demand preferences in nationally preferred innovation designs constitute a national export advantage. One can derive three factors of a national export advantage: domestic demand that is sensitive to the problems and needs of foreign countries, long-time export experience of domestic firms and the similarity of local market conditions to foreign market conditions. Dekimpe et al. (1998) support the hypothesis already suggested by Vernon (1979) that the higher the similarity of cultural, social and economic factors between two countries, the greater the likelihood that an innovation design adopted by one of two countries will be adopted by the other country as well.

The market structure effect focuses mainly on the degree of competition. Competition and entrepreneurial effort has been described as one of the main determinants of international patterns of innovations from Posner (1961) to Dosi et al. (1990). A lead market is commonly a highly competitive market. This is due to the fact that faster development and more market-oriented innovations can be supported by competitive market structures. First of all, buyers tend to be more demanding when they face competition than when they are tightly regulated or hold a monopoly (Porter 1990). Second, competing firms are more under pressure to follow those who have already adopted a new technology (Mansfield 1968). Third, and maybe most importantly, more innovation designs are tested in a competitive market than in a monopolised market. In the absence of barriers to entry or the contestablility of markets (i.e. firms can enter and exit a market, see Baumol et al. 1982) new products and technologies are frequently brought about by new firms (see e.g. Audretsch 1995). This makes the process more efficient in finding the best product by means of search and selection, i.e. the product that is most profitable for the user and thus the dominant design. As a result, a competitive market is more appropriate to find a design that is not only the best within the domestic environment but also the better in each national contexts than the locally selected designs. The more innovative climate of one market compensates for the international differences. This notion of competitive markets as more innovative markets is even found to be correct in the case of Japan's international success, which was long time suspected of being based on protectionism and interventionism: "Its [Japans] industries succeed not when the government manages competition but when it allows competition to flourish." (Porter et al. 2000).

3 Extending the lead market model to eco-innovations

3.1 Peculiarities of eco-innovations: the double externality problem

Applying the lead market model to environmental innovation, certain peculiarities have to be considered. We use the following definition of environmental innovation or eco-innovation (Rennings 2000; Rennings and Zwick 2002): Environmental innovations consist of new or modified processes, techniques, practices, systems and products to avoid or reduce environmental harms. Environmental innovations may be developed with or without the explicit aim of reducing environmental harm. They also may be motivated by the usual business goals such as reducing costs or enhancing product quality. Many environmental innovations combine an environmental benefit with a benefit for the company or user. Eco-innovations produce positive spillovers in both, the innovation and diffusion phase. Positive spillovers of R&D activities can be usually identified for all kinds of innovations. The peculiarity of eco-innovations is that positive spillovers appear also in the diffusion phase due to a smaller amount of external costs compared to competing goods and services on the market. This peculiarity of eco-innovations has been called the double externality problem (Rennings 2000).

Due to be problem of double externalities, eco-innovating firms face the problem that they produce a public good, at least to a certain degree, depending on the character of the good. While for instance biological food creates benefits for both the user (taste, health) and the environment (less pesticides) compared to the consumption of conventional products, the benefits of other environmental goods and services such as electricity from renewable energy have no additional private benefits compared to the use of fossil or nuclear energy. Thus the double externality problem reduces the incentives for firms to invest in eco-innovations. Therefore a need can be identified for measures of environmental and innovation policy. A pure strategy of deregulation and creation of competitive markets is not able to correct these market failures. As long as markets do not punish environmentally harmful impacts and reward environmental improvements, competition between environmental and non-environmental innovation is distorted. This would also mean that only the international diffusion of environmental regulations strengthens the adoption of new environmental technologies abroad, which can therefore be identified as a key success factor of lead markets for environmental innovations.

On the contrary, Porter and van der Linde (1995) argue that even in the absence of policy diffusion strict regulations can put additional pressure on firms to innovate in eco-efficient technologies, and this may improve the competitiveness of domestic firms. The logic behind this is that efficient use of natural resources is at least partly a private good since firms have to pay for the use of water, production of waste etc. Thus natural resource efficiency can be regarded as a part of the total efficiency and competitiveness of a firm. The Porter hypothesis implicitly argues that innovation offsets of environmental policy can occur in two ways:

Case A): Advantages in the short run occur, provided that natural resources are private goods, as in the above mentioned case of organic food that was mentioned above. Another example is the rational use of water, energy and material reducing the bills for electricity, water or waste. In case A the double externality problem is thus not relevant, or even if it is relevant it may be possible to find opportunities for environmental improvements at zero costs. The hypothesis assumes that regulatory pressure triggers firms to develop innovations with win-win-opportunities that they have not seen before due to X-inefficiencies, bounded rationality or incomplete information. While this hypothesis is theoretically valid, it's relevance concerning the magnitude of such win-win-potentials is still controversial (Jaffe et al. 2002).

Case B): Advantages occur in the long run when natural resources have an adequate market price, depending on the international diffusion of environmental policy measures. Case B considers the problem of double externalities. Without policy diffusion the pioneering country has to bear additional costs and a loss of competitiveness.

We assume case B to be typical for eco-innovations. As far as eco-innovations have the character of case A, they do not differ from other innovations. Lead markets for innovations of type A are siumilar to the logic of the market structure effect as described above. The difference in case B is that, here, the incentives to innovate are not generated by competition but by regulation. For such eco-innovations the market structure effect becomes the Porter effect. For the case B innovations the international diffusion of regulation must be added to the lead market model.

3.2 Cross-national policy diffusion

It can be preliminarily concluded that the adoption of national regulations by other countries is one main driver for the international diffusion of environmental innovations. Thus it is important to understand, why environmental regulations from pioneering countries are adopted by other countries and diffuse around the world, ensuring the adoption of the same innovation design internationally. Policy convergence, as discussed among policy analysts, is the frequently observed convergence of policy instruments or policy content, i.e. the institutional tools available for administering policy (Bennett 1991), across administrative borders. We argue in this paper that the motivations of governments to adopt a foreign policy or regulation can be explained by principles similar to the factors governing the diffusion of innovations as presented above.

One process of policy convergence is the cross-national diffusion of policies, the pattern of successive adoption of a policy innovation by imitation. A new policy can be called a policy innovation and the adoption of a specific policy by most countries worldwide an international diffusion of policy innovations (Bennett 1991; Kern et al. 2000). Cross-national policy diffusion can explain the international diffusion of technical innovation designs as well. New regulations of national governments can induce innovations for instance if they require the adoption of new technologies, change relative factors costs or in general change the relative benefits of different designs. If one country introduces a new regulation that induces innovations or spurs the adoption of a specific innovation design, this innovation design will be adopted by users in other countries as well, if other countries introduce this same regulation. The international diffusion of policies in a broad range of fields has already been studied, for example with regard to bureaucratic accountability (Bennett 1997), administrative reform (Peters 1997) and most notably to environmental regulations (see Kern et al. 2000; Jänicke 2000).

Kern et al. (2000) observe the same phenomenon that was to be seen in the international diffusion of innovations: Despite the fact that countries differ in conditions such as ecological problems, requiring different policy instruments, these countries often adopt the same regulations, even down to the wording used by an other country (Bennett 1991). Some countries "sacrifice...autonomy to avoid unnecessary cross-national divergence" (Bennett 1991). If it is more likely that policy makers follow a certain country, this country has an international advantage comparable to the transfer advantage that is discussed in lead market theory. Leading countries are those that are generally more watched by many other countries.

In reviewing the literature of cross-border policy convergence, Bennett (1991) and Dolowitz, Marsh (1996) describe policy internationalisation mechanisms that are analogous to the transfer mechanisms constituting the diffusion of technical innovations. First of all, social problems and policy instruments intended to ease these problems are surrounded by uncertainty. The introduction of regulations offers a test of effectiveness and reduces uncertainty (lesson drawing)[1]. Obviously, countries adopt foreign policies that proved to be effective in one country without harming that country's economy[2]. For instance, deregulation in the telecommunications industry in the United States brought down the price of telephone calls. This led governments of many other countries to deregulation of the telecom sector as well[3].

Policy communities as well as international organisations such as the OECD, the WTO and transnational professional organisations (e.g. aviation, telecommunication, etc.) share experience and have an incentive to harmonise policies among countries, most notably if countries' policies are interdependent, and taking so-called best practises as a guidance. In keeping with the transfer effect of multinational firms, international organisations such as NGOs (Non-governmental organisations) can apply pressure to (or convince) governments worldwide to adopt a specific policy. For instance Greenpeace has transferred the policy of chlorine-free paper worldwide (Sonnenfeld 2000). Furthermore, it has been argued that governments of large countries can force other governments to introduce a specific regulation, for example this international transfer process is suggested to had have happened to in the case of anti-cartel laws in Europe (Majone 1991), deregulation of telecommunication (Hills 1986) or by the regulatory requirements of the IMF for the granting of loans (Dolowitz and Marsh 1996).

It has been observed that some countries' policies are more likely to be copied than others, for instance the United States and Germany (Dolowitz and Marsh, 1996). Yet, political science literature offers little nation-

---

1 "Uncertainty about the cause of problems, the effects of previous decisions or the future causes actors to search for policies they can borrow" (Dolowitz and Marsh 1996).

2 Effective regulation in turn can lead to the fear of political actors in other countries of falling behind its competitors triggering the adoption of the same policies (Dolowitz and Marsh 1996). However, foreign policies not only 'draw lessons' but also legitimate conclusions already reached domestically and are taken to put pressure on the domestic legislation process (Bennett 1991).

3 A detailed but rather disapproving assessment of the international adoption of telecom deregulation policies initiated by the United States is suggested by Hills (1986).

specific characteristics of these regulatory leaders. However, the theoretical incentives behind adopting a similar regulation are already mentioned earlier in this chapter: risk reduction, global externalities and other incentives of multinational organisations to harmonise international conditions and the international reputation and attention a country receives. Thus, a country is more likely to be followed in the adoption of specific policy instruments if it lowers risk most visibly, draws most international attention, spreads political ideas internationally, has more power in international organisations and has a good reputation on a political rather than technical level. As well, and in line with the argument of the next section, the more ideological and institutional similarities a country has to other countries, the more support the international transfer of domestic policies receives (Dolowitz and Marsh 1996).

3.3 Extending the lead market model to environmental innovations

An extension of the lead market model to eco-innovations should consider both the common internationalisation factors of the lead market model and the double externality problem, taking into account the prominent role of policy and it's international diffusion. Figure 3 shows the relationships between the different levels.

Policy patterns consist of instruments such as emission control legislation, tax regimes or subsidies for specific technologies. However, the relationship between politics and innovations is not purely instrumental as most economist want to believe. Policy styles can influence the real effects of instruments (Richardson 1982; Jänicke 2000). For instance, a consens-oriented, calculable and flexible style has been suggested to be more innovation friedndly than other styles (Blazejczak et al. 1999). A policy pattern has a direct influence on the willingness to adopt innovations. This relationship is marked with (1) in the figure. The policy level looks at the likelihood that innovations occur at all. It does not explain under which circumstances an international diffusion of innovations or regulations occurs. Therefore, additional factors have to be considered. The second level of the model constitutes the internationalisation factors. A further analysis should reveal, whether policy styles have an influence on the internationalisation of innovations via the internationalisation of policy instruments (marked as (2) in figure 3). Yet, policy makers are not the only actors on the policy level. Transnational and non-governmental organisations such as the OECD, the United Nations, the WTO, Greenpeace and other pressure groups are spreading environmental policy discussions around the world and facilitate the international harmonisation of policies. These international organisations indirectly and even directly support the standardisation of eco-innovations as well. The export advantage of lead markets that was described above is often constituted by national institutions such as banks, suppliers and customers pressing firms to develop innovations that can be exported. Institutions can have the opposite effect as well. For instance, state-owned monopolies such as telecom and public transport companies often demand nation-specific technologies that do not match the requirements of the world market. Finally, multinational firms - as discussed above - have an incentive to standardise their technology within their global network of affiliates instead of employing different technologies from country to country. Thus they try to push suppliers as well as policy makers to accept – or more often wait for -international agreements on environmental regulations.

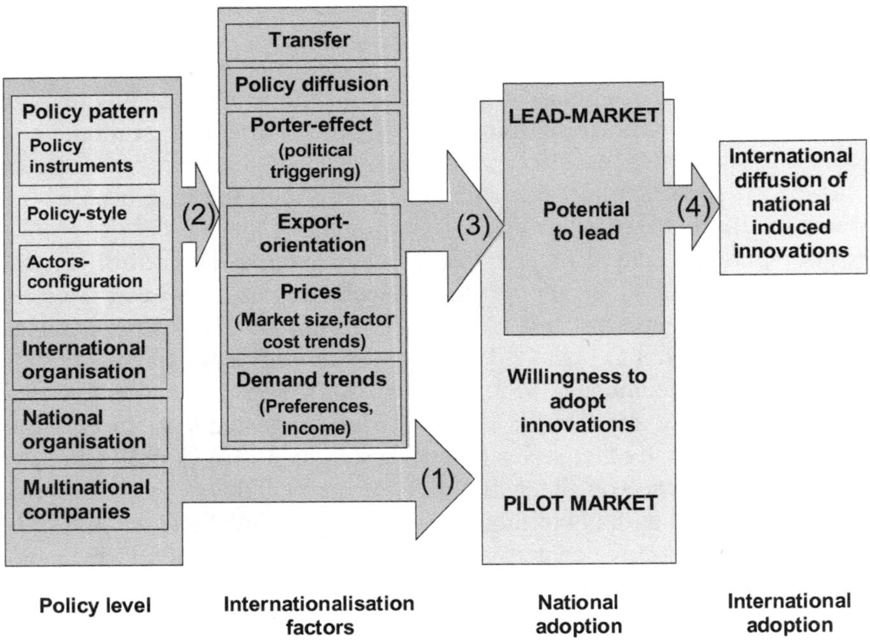

Fig. 3. A framework for analysing the international diffusion of eco-innovations

The discussion on the internationalisation of eco-innovations above has shown that environmental policy-specific arguments can in principle be assigned to the five lead market factors envisioned in innovation economics.

The market structure advantage, however, that focused on competition as the most important factor to push innovations in the conventional lead market model, has been interpreted in the context of environmental innovations as the "Porter effect". In our context the "Porter effect" specifically considers the problem of double externalities (described as case B above). Environmental problems can not be solved simply by deregulation strategies, since the existence of negative external effects requires regulatory measures correcting these market failures. We argue therefore that the market structure advantage shall include the "Porter effect", understood as strict regulation for the respective environmental problem in the lead country. Strict environmental policy can induce innovations, but without policy diffusion the pioneering country has to bear additional costs and a loss of competitiveness in the long run.

The fact that policies are diffusing between nation states or are harmonised in international organisations can provide an additional factor for the internationalisation of innovations. Those countries, that are considered pace makers in the development of environmental policy, do have a transfer advantage. This position might be gained either by innovativeness or a strong position in international organisations.

For the further analysis of lead markets for eco-innovations, the framework as illustrated in figure 3 will be used including both, the modified lead market attributes of countries and the relationship between the policy level and the national adoption of innovations. The traditional impact of policies on the adoption of innovations in a country (1) is not sufficient for making the distinction between lead markets and idiosyncratic national markets. This can be achieved by including the modified lead market factors (2). The analysis within innovation economics has been focused on the relationship between the lead market factors and the potential of a country to lead the adoption of a specific innovation design internationally (3). These factors are likely to improve the chances for an innovation to diffuse internationally (4).

The framework can be used for analysing the effects of policies and actors on the adoption and international diffusion of innovations. Its advantage over former studies on environmental regulations (as e.g. Blazejczak et al. 1999) is that it includes the rationale of an international diffusion of innovations. In the next section we will use the concept of lead markets for two cases, fuel-efficiency of passenger cars and wind energy. In both cases lead markets can be identified that first adopted innovations that could later be exported, and strengthened the competitive advantage of the country in the wind generation and car industry considerably. We discuss what regulations have been used and what were the reasons for the international success of the innovations induced by those.

4 Indicator applications

4.1 Fuel-efficient passenger cars

Fuel-efficiency is a mean to lower the emission of gases that are harmful or cause the greenhouse-effect. Fuel-efficient passenger cars are cars that consume a low level of fuel per 100 km. They are powered by gasoline, hydrogen or they are equipped with both gasoline and electric engines (hybrid cars). In Germany, the most fuel efficient car is the so called "3-Liter-Auto", which means that it consumes less than 4 litres per 100 km. In the 1990s this limit was a realistic goal for most car manufacturers in the context of the European driving habits and design preferences, so that policies, such as favourable tax treatment, were introduced to support it. At the end of the 1990s, there are several German car models that are within this low consumption limit.

Modern fuel-efficient passenger cars use a mix of technologies that are aimed at reducing the fuel consumption of a car. The most effective technologies to reduce fuel consumption are the use of low-weight materials, the sharpening of the aerodynamics of the car body and improvements of the combustion process. The last approach was used most frequently, partly because it is the most efficient, partly because of market preferences. In fact, cars have become even heavier over time and the body design has to follow safety, as well as aesthetic criteria (Franke 1998). Among the motor technologies, the high-pressure direct injection or common-rail injection system were most successful in the 1990s. High-pressure injection improves the combustion, lowers the emission of exhaust gases and at the same time increases the performance, notably the acceleration of cars. In diesel engines the high-pressure injection became almost a standard during the 1990s (figure 4). The modern injection systems were developed by several car companies in Europe and Japan. Germany, however, was the lead market. The US and Japanese markets lagged this technical change, since the share and reputation of diesel powered cars are much lower there (Petersen and Diaz-Bone 1998).

What have been the factors that made fuel-efficiency innovations being adopted in Europe internationally successfully? Europe has the highest fuel prices in the world, making fuel-efficient innovation most beneficial there. Yet, while there is a global trend of increasing gasoline prices, the differences, especially between some European states and the US are still so large, that fuel-efficiency alone cannot persuade US users to adopt the innovation. Only those innovations diffused internationally that not only reduce consumption, but also enhanced other attributes of a car that meet the global demand trend of passenger cars. Another global trend is at work.

Over the time, cars have become more luxurious, heavier and powerful. Fuel-efficient technologies were needed to prevent the consumption levels to increase. In addition to its fuel efficiency feature, high-pressure direct injection proved to be a large improvement for diesel engines that suffered from low acceleration performance. The main reason for the international success was neither the fuel-efficiency legislation, nor the other environmental factors that make fuel-consumption more economical. Fuel efficiency technologies became international successful because they are compatible with the demand preferences in all major markets. Small or micro-compact cars are successful in Japan or parts of Europe, in large countries like the US they probably will never be. High pressure injection is not only used in 3 Liter cars but for all types of cars, even for large luxurious ones. And Diesel engines are more likely to become successful worldwide.

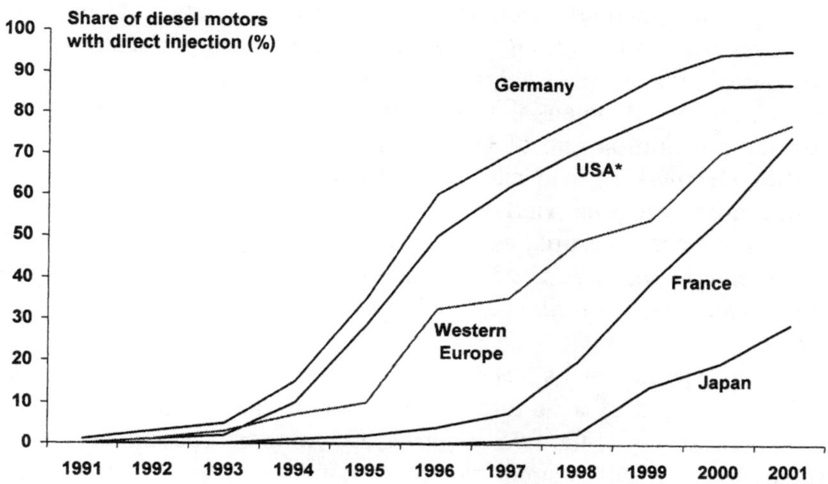

Fig. 4. International Diffusion of Diesel-High-pressure-injection (ZEW, own calculations, data based on interviews with producers); * *predominantly light trucks*

This explains why several innovations failed when they optimised environmental criteria but did not meet other consumer demands with higher priority, such as driving power. A good example is the Golf Ecomatic that was developed by Volkswagen and was introduced to the market in 1993. The Golf Ecomatic switched automatically the motor off when it was not in use, e.g. when the lights are red, starting it again when a gear is used.

This induced a substantial reduction of fuel consumption between 20 and 25 percent, especially in urban traffic. The innovation won several environmental awards, but only 3,000 vehicles were sold in total on the market (Petersen and Diaz-Bone 1998). The driving behaviour of the Gold Ecomatic seemed somehow strange for drivers. Volkswagen introduced another new Golf version nearly at the same time, a Diesel-high-pressure-injection model, being not only fuel–efficient but offering also increased driving-power. The so-called "Golf Turbo Diesel" became very successful.

4.2 Wind energy

The world market for renewable energies and especially wind energy has increased rapidly over the past decades. A driver were the oil crises and the following discussion of environmental impacts of fossil fuels.
The developing wind energy world market is dominated by the small Nordic country of Denmark. Denmark is the pioneer country of electricity production by wind. Although Germany is the country with the largest wind energy capacity installed in the world, the usage of wind energy as a share of total wind potential is still smaller than in Denmark. Figure 5 shows the penetration rate of wind energy use in different countries and identifies Denmark as lead market. Germany follows closely while other countries are developing wind energy with a considerable lag. The penetration rate has been measured as the percentage of exploitation of on-shore wind potential. We have used also other possible indicators, as e.g. the share of wind energy of total electricity production, leading to the same result.

As a consequence of its leading role in using wind energy, Denmark is the largest exporter of wind turbine generators in the world. When import- and export markets of the two countries are compared, it can be seen that Germany exports only a small part of its wind turbines to other countries (BTM-Consult 2002). While Denmark's wind industry is world market oriented, the German wind industry depends more on domestic demand and regulation (Denmark: 81% exports, 19% imports; Germany: 10% exports, 90% imports).

Denmark looks back to a long history with regard to the technical development of wind mills. In 1918 120 Danish energy utilities had already developed a wind mill with a typical size of 20 to 35 kW, providing 3 percent of the total electricity production in Denmark. The so–called "Danish Concept" is traditionally characterised by three rotor blades. Since the fifties direct current generators plants were replaced by generators producing alternating current (modern types of these generators are asynchronous generators). Also the third typical feature of modern wind energy converters was already developed before the oil crisis: today, the converting systems are equipped with pitch or active stall regulations. Both mean different techniques for increasing the flexibility with regard changing wind forces.

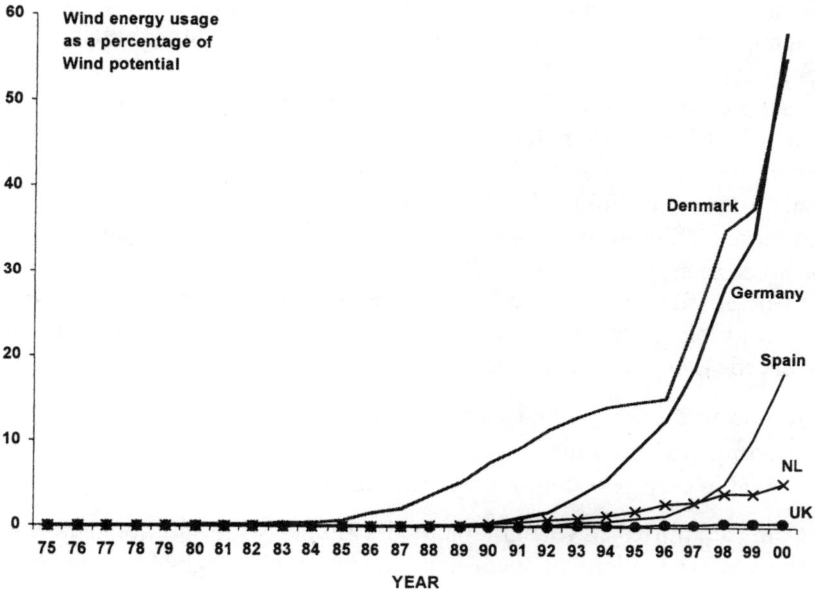

Fig. 5. International diffusion of wind energy; IEA/OECD (2002), Lehmann and Reetz (1995)

While wind energy was seen as being too expensive to compete with fossil fuels, the situation changed during the oil crisis. Several countries began to install big wind power plants with 1 Megawatt and more, like the GROWIAN in Germany. They, however, failed since they were economically inefficient. The main criteria for the technological development of such large wind power plants was their compatibility with the existing system of large, centralised fossil and nuclear energy plants. Energy utilities had no incentives to undermine their own system by developing a decentralised alternative system of renewable energies. In contrast, the technological trajectory of wind energy in Denmark was characterised by more variety and flexibility (Heymann 1995). Although some experiments with large wind power plants were undertaken, the industrial and economic breakthrough was achieved by the continuos improvement of smaller converters. The 55 Kilowatt generation of wind turbine generators, being developed between 1980 and 1981, realised cost reductions of about 50 percent. In the 80s many technology support programmes were set up all over the world, e.g. in California. Thousands of Danish Micon 55 Kilowatt wind turbines were exported to Palm Springs. The Danish producers had the first mover advantage. They had started with the industrial production of wind turbines five years earlier than their competitors (Danish Wind Industry Association 2002). Since the 80s the size of wind turbines is increasing continuously. The diameter of new rotor blades increased from 23 meters in 1990 to 44 meters in 1993 and 63 meters in 1997. Modern generators have already passed the Megawatt class of the GROWIAN generation. In the year 2000, 15 plants had an installed capacity of 2 Megawatt and more. Offshore generators are planned with 3 to 5 megawatt (Institute for Solar Energy Technology 2002).

Policies for wind energy rather varied from country to country. In Europe, three different types of strategies supporting wind energy can be observed (Langraf and Kellner 2000; Haas 2001; BTM-Consult 2002):

- Renewable Energy Feed Tariffs (REFITs),
- Bidding systems and
- Tradable permit system for renewables.

Some countries have implemented systems with additional incentives, such as tax reductions or specific depreciation privileges for renewable energies.

Substantial differences can be identified when the regulation systems are related to development of a national wind industry (Haas 2001). In countries with REFITs system wind industry developed rapidly, for instance in Denmark, Germany, Italy, Austria and Spain, until 1994 and 1995 also in Ireland and the Netherlands. In countries with bidding systems wind energy use developed very slowly, independent of the existing wind resources. In France, United Kingdom and Ireland wind industry is poorly developed although coastal regions with high wind potentials exist.

Finally, the system of tradable permits for green electricity is still too young to evaluate the impact on the wind industry. It can be expected that the success will depend heavily on the design of the trading system and of the underlying quota system. The European Commission has formulated the ambitious goal of doubling the share of renewable energy within 10 years until 2010 (Jung 2002). If the system of tradable quotas is based on such ambitious goals, it may lead to a very dynamic development of wind industry and wind energy technologies. If the quotas are less ambitious, the development of wind industry may break down. Denmark has experienced such a break down in 2001 after switching from the REFITs system to the quota system. In 2001 only 117 MW new wind capacity were installed in Denmark, compared to 802 MW in Spain and 2,659 MW in Germany (BTM Consult 2002).

5 Conclusions

In the final section we draw some conclusions concerning the relevance of our identified lead market factors for the two cases.

Price advantages seem to be a relevant but not the dominating driver of the international diffusion of the innovation in both cases. In the case of fuel-efficient cars, high prices for small 3 litre cars are still a substantial bottleneck for their success. Lower prices for Diesel compared to petrol in Germany and France has increased the market share of Diesel cars in these countries, reaching a share of 25 and 50 percent respectively. This is, however, no global trend yet. In the case of wind energy it can be stated that renewable energies are still more costly compared to conventional energy sources such as fossil fuels or nuclear energy. Since renewable energies produce fewer external effects, regulatory measures are needed for internalisation. Experience shows that REFITs systems were most successful concerning internalisation of external effects (subsidies for environmental friendly energy sources), production of renewable energy at reasonable prices and development of a competitive domestic wind industry. Obviously protected niche markets are needed at least temporarily to create an attractive environment for investments into alternative energy plants. In contrast, bidding systems led to cost reductions but also to uncertain and risky investment conditions. No country with bidding systems could reach an substantial share of the wind energy world market. The different experience of policies leads to a policy convergence favouring the REFITs system as the most successful and globally dominating policy.

Demand advantages are crucial for the lead market of fuel-efficient cars since other criteria of global demand are still more important than environmental criteria. People demand fuel-efficient cars only if they meet performance criteria additionally to ecological criteria with no, low or even negative costs. Negative costs can occur due to lower fuel consumption of eco-efficient cars. Thus fuel-efficient cars can be subsumed under Case A as formulated in the section on factors of environmental lead markets. This explains the success of the Diesel-High-pressure-direct-injection technology. Reductions of fuel consumption over the past 30 years have nearly been compensated by heavier and more powerful cars. Since there is no real world market for small 3 litre cars, especially not in the US, only innovations that offer additional benefits and anticipate consumer's demand trends diffuse world-wide. Fuel-efficient technologies are employed not only in small or micro compact cars, the prototype of a fuel efficient car, but also in large luxurious cars such as Mercedes-Benz. Those cars are internationally more successful than any other type and it can be expected that the most successful fuel-efficient car will rather be a midsize sedan than the 3-litre cars currently offered. In the case of wind energy consumer's demand trends are less important. Electrical energy is a homogeneous product and most consumers do not care which energy source they get the electrical energy from (a typical example of Case B as described in the section on factors of environmental lead markets). It is more important for lead markets to anticipate trends of global environmental problems. Denmark as the lead market of wind energy has anticipated this global trend towards alternative energy resources early. The context of the Danish market then facilitated the development of energy generation that could be exported.

Transfer advantages can be identified in both cases since the R&D activities of the German automobile firms and the respective efforts of the Danish wind industry are intensively watched by other countries. The innovations have to prove their feasibility and practicability in daily life before they diffuse internationally. In the fuel efficiency case it is an obstacle for radical innovations like hybrid cars or fuel cells to prove practicability because they require substantial changes of driving habits or infrastructure etc. Thus incremental innovations as the Diesel-high-pressure-direct-injection are preferred. Geographical and regulatory characteristics of the US automobile market are an obstacle for small 3 litre cars and Diesel engines. In the case of renewable energies radical innovations like wind energy have somehow better chances since they do not require any changes of consumer behaviour. Bottlenecks are higher costs and compatibility with the existing energy system (including infrastructure and conflicts with lobbyists of conventional energy sources).

Export advantages address the similarity of market conditions at home and abroad. They are more important in the wind energy case than for fuel-efficient cars. Wind turbine technologies are very similar all over the world, creating substantial transfer advantages for exporting countries. Denmark produced a large share of the wind turbines which were installed in the US. They were identical with the domestic installations.

Finally the market structure or Porter effect has proved to be relevant in both cases. In the case of wind energy strict regulation, together with an anticipated regulatory trend as described above, can be seen as the dominating success factor for Denmark as a lead market. Without strict regulations and international policy diffusion renewable energies would not be competitive. For fuel efficient cars the Porter effect is less important since environmental regulation is to date still outweighed by consumer preferences that steer diametrically into the opposite direction.

Summing up, all lead market factors seems to be at least relevant for environmental innovations. The importance of the Porter effect depends on its relation to global demand and regulatory effects. If national regulation is supported by global demand or regulatory trends, a strong effect can be identified, as was shown in the cases of wind energy in Denmark and Diesel-high-pressure-direct-injection in Germany. If it is not supported, the market remains idiosyncratic, as could be seen in the failure of the the Golf Ecomatic.

The Innovation Directorate of European Commission has "proposed to further investigate the parameters involved in the formation of lead markets, including examination, together with industrial representatives, of the potential for specific industrial sectors to benefit from European lead markets as a step towards a stronger presence on the international market" (ITT 2003). This proposal can also be supported for lead markets of environmental innovations.

6 Acknowledgements

This paper results from the research project "Policy-Frameworks for the Development of International Markets for Innovations of a Sustainable Economy – from Pilot Markets to Lead Markets" funded by the German Federal Ministry of Education and Research (BMBF) under the research initiative "Sustainability and Innovation – Frameworks for Innovation towards Sustainability". We are grateful for helpful comments from our project partners Martin Jänicke, Klaus Jacob, Dietmar Edler, Jürgen Blazejczak, Ulrich Petschow and Thomas Loew, and for technical support from our research assistants Ilja Karabanow and Sarah Gramm.

7 References

Audretsch D (1995) Innovation and Industry Evolution. Cambridge

Baumol W, Panzar J, Willig R (1982) Contestable Markets and the Theory of Industry Structure. New York

Beise M (2001) Lead Markets: Country-specific Success Factors of the Global Diffusion of Innovations, ZEW Economic Studies Vol. 14. Heidelberg

Bennett CJ (1991) Review Article: What is Policy Convergence and What Causes it? British Journal of Policy Studies 21: 21-233
Bennett CJ (1997) Understanding Ripple Effects: The Cross-National Adoption of Policy Instruments for Bureaucratic Accountability. Governance 10(3): 213-233
Blazejczak J, Edler D, Hemmelskamp J, Jänicke M (1999) Environmental Policy and Innovation – An International Comparison of Policy Patterns and Innovative Impacts. In: Klemmer P (ed) Innovation and the Environment. Berlin
BTM Consult (2002) BTM statistics. http://www.btm.dk
Dekimpe MG, Parker PM, Sarvary M (1998) "Globalisation": Modelling Technology adoption Timing across Countries, INSEAD working paper No. 98/69/MKT
Danish Wind Industry Association (2002) Windpower Information. http://www.windpower.dk
Dolowitz D, Marsh D (1996) Who Learns What from Whom: a Review of the Policy Transfer Literature. Political Studies 44: 343-357
Dosi G, Pavitt K, Soete L (1990) The Economics of Technical Change and International Trade. New York
Franke A (1998) Auf dem Weg zum 3-Liter-Auto – Minderung der verkehrsbedingten CO2-Emissionen. Abschätzung der innovativen Wirkungen umweltpolitischer Instrumente – dargestellt am Beispiel des Systems Straßenverkehr. München
Franko LG (1976) The European Multinationals. London
Haas R (ed) (2001) Promotion Strategies for Electricity from Renewable Energy Sources. Institute of Energy Economics. Wien
Heymann M (1995) Die Geschichte der Windenergienutzung 1890-1990. Frankurt
Hills J (1986) Deregulating Telecoms: Competition and Control in the United States, Japan and Britain. London
Institute for Solar Energy Technology (ISET) (2002) Renewable Energy Information System on Internet – REISI. http://reisi.iset.uni-kassel.de
ITT (Innovation and Technolgy Transfer) (2003) Current Challenges for EU Innovation Policy. Stimulate greater market dynamisn and exploit the concept of lead markets. ITT, the newsletter of the European Commission's Innovation Directorate
IEA/OECD – International Energy Agency and OECD (2002) Energy Balances of OECD Countreis 1960 – 2000. Paris, International Energy Agency Statistics, OECD
Jaffe AB, Newell RG, Stavins RN (2002) Environmental Policy and Technological Change. Environmental and Resource Economics 22: 41-69
Jänicke M (2000) Ökologische Modernisierung als Innovation und Diffusion in Politik und Technik: Möglichkeiten und Grenzen eines Konzepts. Forschungsstelle für Umweltpolitik Report 00-01
Jung A (2002) Rahmenbedingungen für die Nutzung erneuerbarer Energien (Paper presented on the Conference "Renewable Energies" in Wiesbaden)

Kalish S, Mahajan V, Muller E (1995) Waterfall and Sprinkler New-Product Strategies in Competitive Global Markets. International Journal of Research in Marketing 12: 105-119

Kern K, Jörgens H, Jänicke M (2000) Die Diffusion umweltpolitischer Innovationen: Ein Beitrag zur Globalisierung von Umweltpolitik. Zeitschrift für Umweltpolitik & Umweltrecht 4: 507-546

Langraf B, Kellner T (2000) Windenergie in Europa: Was bringen Gesetzgebung und Fördermaßnahmen. Interwind AG im Auftrag des Bundesamtes für Energie (Ch) Zürich

Lehmann H, Reetz T (1995) Zukunftsenergien – Strategien einer neuen Energiepolitik. Basel

Levitt T (1983) The Globalisation of Markets, Harvard Business Review 61(3): 92-102

Majone G (1991) Cross-National Sources of Regulatory Policy Making in Europe and the United States. Journal of Public Policy 11: 79-106

Mansfield E (1968) Industrial Research and Technological Innovation: An Econometric Analysis. New York

Mowery D, Nelson R (1999) Introduction. In: Mowery D, Nelson R (eds) Sources of Industrial Leadership. Cambridge

Peters G (1997) Policy Transfer Between Governments: The Case of Administrative Reform. West European Politics 20: 71-88

Petersen R, Diaz-Bone H (1998) Das Drei-Liter-Auto, Berlin

Porter M, Takeuchi H, Sakakibara M (2000) Can Japan Compete? London

Porter ME, van der Linde C (1995) Towards a New Conception of the Environment–Competitiveness Relationship. Journal of Economic Perspectives 9(4): 97-118

Porter M (1990) The Competitive Advantage of Nations. New York

Posner MV (1961) International Trade and Technical Change. Oxford Economic Papers 30: 323-341

Rennings K (2000) Redefining Innovation – Eco-Innovation Research and the Contribution from Ecological Economics. Ecological Economics 32: 319–332

Rennings K, Zwick T (2002) The Employment Impact of Cleaner Production on the Firm Level – Empirical evidence from a Survey in Five European Countries. International Journal of Innovation Management (IJIM), Special Issue on „The Management of Innovation for Environmental Sustainability“6(3): 319–342

Richardson J (ed) (1982) Policy Styles in Western Europe. London

Sonnenfeld DA (2000) Contradictions of Ecological Modernisation: Pulp and Paper Manufacturing in Southeast Asia. In: Mol A, Sonnenfeld D (eds) Ecological Modernisation Around the World: Perspectives and Critical Debates. London

Takada H, Jain D (1991) Cross-National Analysis of Diffusion of Consumer Durable Goods in Pacific Rim Countries. Journal of Marketing 55: 48-54

Tilton JE (1971) International Diffusion of Technology: The Case of Semiconductors. Washington

Utterback JM (1994) Mastering the Dynamics of Innovation. Boston

Vernon R (1966) International Investment and International Trade in the Product Cycle. Quarterly Review of Economics 88: 190-207
Vernon R (1979) The Product Cycle Hypothesis in a New International Environment. Oxford Bulletin of Economics and Statistics 41(4): 255-267

Policy Impacts on Macroeconomic Sustainability Indicators when Technical Change Is Endogenous[1]

Joachim Schleich, Rainer Walz, Bernd Meyer and Christian Lutz

1 Objective

The objective of this paper is to link the concepts developed in the working group "Indicators for sustainable innovation" (see the introduction of Horbach in this volume) to the project „Innovationen und Luftschadstoffemissionen - Eine gesamtwirtschaftliche Abschätzung des Einflusses unterschiedlicher Rahmenbedingungen bei expliziter Modellierung der Technologiewahl im Industriesektor (Schleich et al. 2002a). In particular, the determinants of innovations considered in this project, the indicators used for the description of innovation and the impact dimensions of innovations which can be analysed with the proposed modelling approach are described. This contribution differs from the other contributions in this volume in various ways. First, analyses are based on econometric estimations, rather than case studies. Second, indirect effects such as price- substitution- and income-effects are included. And third, macroeconomic implications such as the impacts on employment or GDP can be analysed.

2 Introduction

The economic and ecological effects resulting within the frame of policy simulations in environmental-economic models are decisively influenced by the modelling of technological progress. For example, in models calculating the costs of climate protection, varying model results can, among other things, be traced back to varying assumptions about the development

[1] The project is sponsored by the German Federal Ministry of Education and Research (BMBF) in the funding program "Frameworks for Innovation towards Sustainability (RIW)" and is accompanied by the project sponsor Environmental and Climate Research in Munich.

of technological progress in the baseline scenarios[2]. In these models, technological progress usually is represented through a trend variable. That is, endogenous, policy-induced technological progress as described by Hicks (1932) is not represented[3,4]. Even in models allowing for endogenous technical change there is usually no link to the actual technologies which are responsible for the technological development. And yet in neo-classical general equilibrium models, production functions exhibiting a nonzero constant elasticity of substitution (CES) are explicitly assumed for the individual branches and cost-minimising factor demand functions derived from them (Welsch 1996; Böhringer 1999). The changes in factor price relations triggered by price or other instruments result in substitution processes, but technical change remains exogenous. As a consequence, in these models, policy interventions have no impact on technological change.

Further criticism with reference to the portrayal of technological change in computable general equilibrium models is sparked off by the postulated type of neo-classical production functions. These typically imply unlimited factor substitution possibilities which ceteris paribus result in an underestimation of the costs of climate policies. In particular the large industrial "energy consumers" such as the electricity industry, steel production, producers of non-ferrous metals, the cement industry or paper manufacturing are better characterised by limited production relations of the "putty-clay" type in the factor substitution possibilities.[5] For production functions of this type, a choice can be made when making investment decisions between different limitational processes, whereas the input structures of the existing plants can no longer be altered.

To summarise, it can be concluded that, first of all, innovation and technological change are only represented superficially in the predominant

---

2 See, for example, Weyant (1993) or, more recently, Jaffe et al. (2003). Other reasons for differing results include the level of aggregation, the technologies available in the baseline, or differ-ences in expectations about future prices and other economic developments (Forum für Energiemodelle und Energiewirtschaftliche Systemanalysen in Deutschland 1999).

3 See Jaffe and Stavins (1994a). Exceptions include Carraro (1998), Nordhaus (1999), Goulder and Schneider (1999), Goulder and Mathai (2000) or Buoananno et al. (2003). For a comprehensive survey on technological progress in environmental-economic models see Löschel (2002).

4 A different form of endogenous technological progress which is not examined here results from so-called learning-by-doing effects, which in normative work imply early investments in reduction measures (Van der Zwaan et al. 2002).

5 The share of putty-clay technologies in total industrial production is estimated at 50 % to 70 % - even higher in energy-intensive sectors (Gilchrist and Williams 2000)

models (Frohn et al. 1998; Hemmelskamp 1999; FIU 1996), and that, secondly, the assumption of complete factor substitution does not correctly reflect the actual production processes in many production sectors.

3 Modelling approach

In this project, a new modelling approach is presented which takes up the challenge of these criticisms in two ways. In an integrated bottom-up/top-down approach based on the example of crude steel production in Germany, it is first demonstrated how technical progress can be portrayed as process-related and policy-induced. Second, it is shown how technology choice between limitational processes can be explicitly modelled and implemented in the econometric input-output model PANTA RHEI[6]. To do so, among others, investments, production amounts, detailed input structures and the process-specific input demand of the respective best-practice technologies (trajectories)[7] are determined for the historical observation period (1980-1996) for the different process lines (paradigms) (Dosi 1982, 1988). Based on these data, the paradigm-specific investments, i.e. the choice of technology and the development of technical change in the model can be estimated econometrically as a function of prices and other variables. The correlations found then serve as the basis for the future-oriented policy simulations.

In general, the innovation process distinguishes five stages: recognition, invention, development, implementation, and diffusion (Modesto 1980; see the contribution of Horbach in this volume). The focus of the new modelling approach developed in the project is on development and diffusion, which will be described in more detail below.

The new modelling approach has been developed for three case studies. The following sections describe the indicators used for the determinants, descriptions and impacts according to one of these case studies, the steel sector.

---

6 For a short model overview see the appendix. For more details see Meyer et al. (1999) or Lutz (2000). A description of the new modelling approach for the steel sector together with simulation results can also be found in Lutz et al. (2004).

7 A detailed documentation of the procedure of data generation and the estimates conducted are found in Schleich et al. (2002b).

4 Determinants of innovation

Starting from the best-practice trajectories of the technological paradigms it is possible to endogenise technical progress in the model. Technical measures to improve energy efficiency which are reflected in the chronological changes of the energy input structures include, among others, decreasing the consumption of reducing agent in iron making - e. g. by partly substituting coke with injected pit coal, fuel oil or scrap plastics - measures in integrated ironworks, in coking plants and in sintering plants as well as control technology measures and the optimisation of the energy supply in electric arc furnace steel works. Thus, while innovations can generally be classified as product and process innovation, organisational and institutional innovation, in the project, technological progress is process-integrated.

To analyse the determinants of technological progress, the correlation between the development over time of the best-practice energy consumption of the electric or oxygen steel production respectively and a number of price variables as well as the R&D expenditure of the steel industry and the mechanical and electrical engineering sectors is econometrically estimated. The hypothesis behind including the R&D spending of the mechanical and electrical engineering sectors is that the producers of investment goods target their research efforts within the scope of the given respective paradigm to offering where possible those investment goods which minimise the production costs in the demand sectors (see Erdmann 1993)[8]. Thus, demand for environmental innovations which translates into improved environmental / energy performance is primarily a by-product of companies' efforts to reduce production costs (cost-push hypothesis). When developing new technologies, the suppliers of energy-efficient technologies take into account the cost-effects of their products for their customers. Besides economic determinants, other variables were included as proxies to test the statistical significance of other "barriers" on the development of technical progress. Specifically, it was analysed whether the data provide empirical support for the X-inefficiency hypothesis: X-inefficiencies from lack of competition in the production of steel may result in too little technological progress. For the development of specific energy consumption, the Herfindahl-index of industry concentration and the share of imported steel to domestic production (impact of market structure) were used as proxies for (potential lack of) competition in the regression equation. However, neither variable turned out to be statistically sig-

[8] This assumption is supported, for example, by the results of the empirical work of Grupp (1999).

nificant. To test the significance of other barriers such as information and other transaction costs, bounded rationality or asymmetric information (Brown 2001; DeCanio 1993; Eyre 1997; Howarth and Andersson 1995; Jaffee and Stavins; 1994a, 1994b; Ostertag 2002; Simon 1947; Sorrell et al. 2004; Stern 1986) or other, so-called „soft-context factors“ (Klemmer et al. 1999) adequate proxies and sufficient data on these proxies would be required. In general though, the results are consistent with the hypothesis that the companies are to be considered as "supplier-dominated firms" with regard to the characterisation of technical change (Pavitt 1984).

The realisation of technical progress occurs in the model primarily through the diffusion of best-practice capital goods and takes place primarily through the technology choice of the investment decision (Dosi 1988). These investments in turn are estimated econometrically as a function of the paradigm-specific outputs, the existing production capacities, relative input prices, and of the real interest rate.

In the new modelling approach, policy interventions can play an important role for the development and the diffusion of new technologies. In particular, price policies, such as energy or CO2 taxes, or an international emissions trading system will in-crease the costs of energy which in turn induces the technology supply sectors, that is mechanical engineering and electrical engineering to develop less energy or CO2-intensive technologies. As described above, these "cleaner" technologies will then enter the capital stock when new investments are made. Likewise, the effect of higher R&D efforts could be modelled. Thus, modelling technological progress as endogenous, policy changes trigger changes among the other determinants of innovations which in turn lead to the innovations (inventions) themselves. Unlike most other approaches, the modelling approach presented allows for a process-specific analysis of the impacts of changed frame conditions, the effects of which on the choice of technology on the one hand and the technological progress on the other can be described endogenous to the model.

For steel production, such analyses were carried out for two policy simulations – an international CO2 tax and a decrease in the price of scrap – which affect input prices in the production of oxygen versus electric arc furnace steel. Other types of simulations are conceivable as well. If better data on the capital stock were available (vintage model), "windows of opportunity" could be explored, that is, the exploitation of time slots resulting from investment cycles (Erdmann 1999; Zundel et al. 2003; Zundel et al. 2004). Likewise, the influence of soft factors, such as policy credibility, could be analysed, or the impact of intersectoral structural change such as changes in demand for steel-purchasing sectors could be incorporated.

In summary, the following types of determinants for technical progress and its diffusion are used:

- various price variables
- R&D expenditure of case study industry and the supplying industries (e.g. mechanical and electrical engineering)
- market structure (industry concentration, share of imports)
- output and existing production capacities,
- relative input prices for case study sector,
- real interest rate (reflecting the costs of capital), and
- policy changes described as scenarios which are simulated.

Within this project, the impact of these determinants is assessed based on econometric estimations, which rely on time series of historic data at the sectorial level.

5 Description of innovation

The modelling approach uses the concept of technological paradigms (Dosi 1982, 1988). Within each technological paradigm, incremental innovations occur along a trajectory. At the same time, innovation can take place by substituting one paradigm for the other. For each paradigm, the trajectories are described according to a technology-based description, with a focus on energy inputs.

One of the case studies for which the model was developed is the steel sector. For steel production in Germany two relevant production lines can be distinguished:

- oxygen steel production, i. e., the process of producing primary materials following the route sintering plant (ore concentration) / coking plant - blast furnace (iron making) - converter (steel production), as well as
- electric arc furnace steel production, i. e. the process of producing secondary materials primarily in electric arc furnaces (to a lesser extent in induction furnaces) based on smelted down scrap.[9]

---

9 Scrap is also used (in small amounts) in oxygen steel production to regulate the temperature of the exothermic conversion process in the converter. Primary materials can also be used when producing electric arc furnace steel. One example, which is practised at only one location in Germany, is the direct reduction using natural gas of iron ore to sponge iron (DRI direct reduced iron), which is used in the electric arc furnace steel process.

From an energetic perspective, the production of electric arc furnace steel is more attractive, since it requires less than half the primary energy demand of the blast furnace-oxygen steel route. The development of the various production lines in Germany is presented in figure 1.

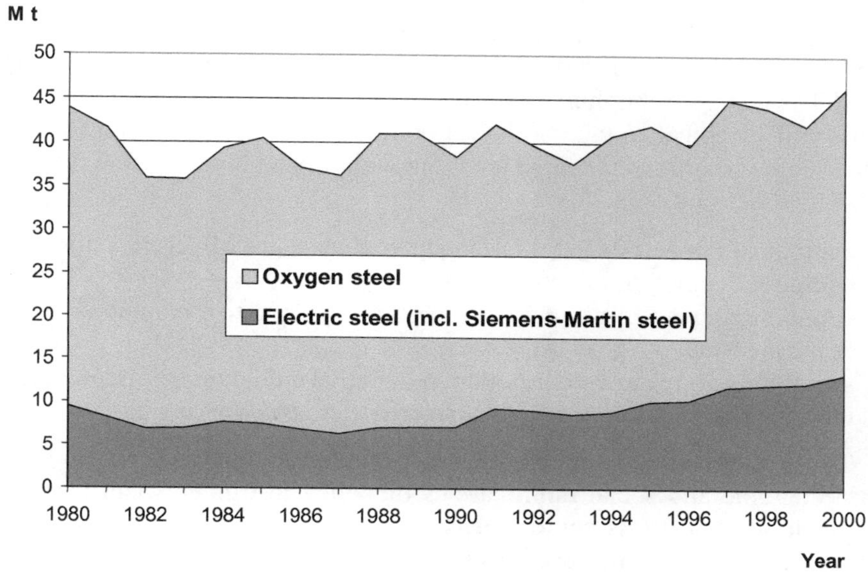

Fig. 1. Process lines of crude steel production; *Wirtschaftsvereinigung Stahl/VDEH, various volumes*

The actual technological conditions are explicitly portrayed in the form of the most important input coefficients for both main production lines of electric arc furnace steel and oxygen steel. They describe the innovations along the trajectories of the given paradigms. Thus, the model portrays continuous technical progress endogenously within the given technical paradigms, but basic innovations, i.e. the invention of paradigms are given exogenously. Within the overall modelling framework which consists of an input-output matrix of all the sectors of the economy every innovation also leads to changes in the input structure of the process considered. Thus, the description of innovation is not restricted to the sector, but leads to a description of innovation at a rather aggregated macro level. Basically, the innovations in the economy are measured by the changes of all input coefficients in the sectors. Thus, the innovations are measured by changes of a predefined matrix.

6 Ecological, economic and social policy impacts

The integrated bottom-up/top-down approach allows for both, the process-specific analysis of policy impacts on the diffusion and the development of technological processes, as well as the analysis of macroeconomic effects. Unlike in most other contributions in this volume, the analyses are based on econometric estimations, which are at the core of the PANTA RHEI model as well as technology-specific estimates of the development and diffusion of technical change. Within the model, the effects of climate and other policy instruments can be analysed with respect to changes in the following dimensions:

- ecological impact: CO_2–Emissions (greenhouse gas effect), energy consumption (depletion of non-renewable resources);
- economic impact: gross domestic product, quantitative employment effects, investments (growth), government budget;
- social impact: income distribution, qualitative employment effects[10] (job qualification requirements, job characteristics, and working hours)

Thus, in the sense of a comprehensive understanding of sustainability, the economic and social ramifications of policy instruments can be analysed in addition to its environmental consequences in an integrated and consistent policy framework. However, no attempt is made to evaluate (weigh) the trade-offs between the different dimensions.

7 Conclusion

This contribution differs from the case-study approach in various ways. First, analyses are based on econometric estimations, rather than case studies. On the one hand, this allows for hypothesis testing based on a larger set of observation and for generalisations of the results where the focus is on the sectoral level. On the other hand, the impact of decision- and organisational processes in companies on innovations cannot be captured. Likewise, the analyses are restricted to factors which can be included as variables in a model, and where sufficient data is available. In addition, since the modelling approach relies on historic data and historic structures, Second, indirect effects such as price- substitution- and income-effects are included. And third, macroeconomic implications such as the impacts on

[10] These qualitative employment effects are analysed by linking the model results with data from the lates microcensus survey for Germany.

employment or GDP can be analysed. In that sense, the modelling approach applied not only allows for the process-specific analyses of policy impacts on the diffusion and the development of technologies. It also allows for the analyses of policy impacts on indicators for the environmental, economic and social ramifications within an integrated and consistent framework.

8 References

Böhringer C (1999) Allgemeine Gleichgewichtsmodelle als Instrument der energie- und umweltpolitischen Analyse. Theoretische Grundlagen und empirische Anwendung. Frankfurt

Brown MA (2001) Market failures and barriers as a basis for clean energy policies. Energy Policy 29: 1197-1207

Buoananno P, Carraro C, Galeotti M (2003) Endogenous Induced Technical Change and the Costs of Kyoto. Resource and Energy Economics 25: 11-35

Carraro C (1998) Climate Modelling and Policy Strategies. The Role of Technical Change and Uncertainty. Energy Economics 20: 463-471

DeCanio SJ (1993) Barriers within firms to energy efficient investments. Energy Policy 21: 906-914

Dosi G (1982) Technological Paradigms and Technological Trajectories: A Suggested Interpretation of the Determinants and Directions of Technical Change. Research Policy 11: 147-162

Dosi G (1988) The nature of the innovative process. In: Dosi G, Freeman C, Nelson R, Silverberg G, Soete L (eds) Technical Change and Economic Theory. London and New York

Erdmann G (1993) Elemente einer evolutorischen Innovationstheorie. Tübingen

Erdmann G (1999) Zeitfenster beachten. Möglichkeiten der Ökologisierung der regulären Innovationstätigkeit. Ökologisches Wirtschaften 2: 21-22

Eyre N (1997) Barriers to energy efficiency: more than just market failure. Energy & Environment 8(1): 25-43

FIU (1996) Der Einfluss von Energiesteuern und -abgaben zur Reduktion von Treibhausgasen auf Innovation und technischen Fortschritt - Clearing Studie. Essen

Forum für Energiemodelle und Energiewirtschaftliche Systemanalysen in Deutschland (ed) (1999) Energiemodelle zum Klimaschutz in Deutschland. Strukturelle und gesamtwirtschaftliche Auswirkungen aus nationaler Perspektive. Heidelberg

Frohn J, Leuchtmann U, Kräussl R (1998) Fünf makroökonometrische Modelle zur Erfassung der Wirkungen umweltpolitischer Maßnahmen - eine vergleichende Betrachtung. Band 7 der Schriftenreihe „Beiträge zu den umweltökonomischen Gesamtrechnungen“. Stuttgart

Gilchrist S, Williams J (2000) Putty-Clay and Investment: A Business Cycle Analysis. Journal of Political Economy 108: 928-960

Goulder LH, Mathai K (2000) Optimal CO2 Abatement in the Presence of Induced Technological Change. Journal of Environmental Economics and Management 39: 1-38

Goulder LH, Schneider S (1999) Induced Technological Change, Crowding Out, and the Attractiveness of CO2 Emissions Abatement. Resource and Environmental Economics 21(3-4): 211-253

Grupp H (1999) Umweltfreundliche Innovation durch Preissignale oder Regulation? Eine empirische Analyse für Deutschland. Jahrbücher für Nationalökonomie und Statistik 219 (5+6): 611-631

Hemmelskamp J (1999) Umweltpolitik und technischer Fortschritt. Schriftenreihe des Zentrums für Europäische Wirtschaftsforschung. Heidelberg

Hicks J (1932) The Theory of Wages. London

Howarth RB, Andersson B (1993) Market barriers to energy efficiency. Energy Economics 15: 262-272

Jaffe AB, Stavins RN (1994a) Energyefficiency investments and public policy. The Energy Journal 15(2): 43-65

Jaffe AB, Stavins RN (1994b) The energyefficiency gap: What does it mean? Energy Policy, 22: 804-810

Jaffe AB, Newell RG, Stavins RN (2003) Technological Change and the Environment. In: Mäler K-G, Vincent JR (eds) Handbook of Environmental Economics vol 1. Amsterdam

Klemmer P, Lehr U, Löbbe K (1999) Umweltinnovationen - Anreize und Hemmnisse. Berlin

Löschel A (2002) Technological Change in Economic Models of Environmental Policy. A Survey. Ecological Economics 43: 105-126

Lutz C (2000) NOx Emissions and the Use of Advanced Pollution Abatement Techniques in West Germany. Economic Systems Research 12: 305-318

Lutz C, Meyer B, Nathani C, Schleich J (2004) Endogenous technological change and emissions: the case of the German steel industry. Energy Policy 32, forthcoming

Meyer B, Bockermann A, Ewerhart G, Lutz C (1999) Marktkonforme Umweltpolitik. Wirkungen auf Luftschadstoffemissionen, Wachstum und Struktur der Wirtschaft. Heidelberg

Modesto MA (1980) Entrepreneurs, Champions and Technological Innovation. Sloan Management Review 21 (2):59-76

Nordhaus W (1999) Modelling Induced Innovation in Climate-Change Policy. (Paper presented at the Workshop on Induced Technological Change and the Environment. IIASA, Laxenberg, Austria)

Ostertag K (2002) No-Regret Potentials in Energy Conservation: An Analysis of their Relevance, Size and Determinants. Heidelberg

Pavitt K (1984) Sectoral patterns of technical change: Towards a taxonomy and a theory. Research Policy 13: 343-373

Schleich J, Meyer B, Lutz C, Nathani C, Schön M (2002a) „Technologiewahl, technischer Fortschritt und Politiksimulationen -Ein neuer Modellierungsansatz am Beispiel der Stahlerzeugung; Zwischenbericht zum BMBF-Projekt: "Innovationen und Luftschadstoffemissionen - Eine gesamtwirtschaftliche

Abschätzung des Einflusses unterschiedlicher Rahmenbedingungen bei expliziter Modellierung der Technologiewahl im Industriesektor. Karlsruhe and Osnabrück

Schleich J, Nathani C, Ostertag K, Schön M, Walz R, Meyer B, Lutz C, Distelkamp M, Hohmann F, Wolter M-I (2000b) Innovationen und Luftschadstoffemissionen - Eine gesamtwirtschaftliche Abschätzung des Einflusses unterschiedlicher Rahmenbedingungen bei expliziter Modellierung der Technologiewahl im Industriesektor; Dokumentation Stahlindustrie. Karlsruhe and Osnabrück

Simon H (1947) Administrative Behaviour: A Study on Decision-making Processes in Administrative Organizations. New York

Sorrel S., Schleich J, O'Malley E, Scott S (2004) The Economics of Energy Efficiency: Barriers to Cost-Effective Investment. Cheltenham

Stern PC (1986) Blind spots in policy analysis: What economics doesn't say about energy use. Journal of Policy Analysis and Management 5(2): 200-227

Van der Zwaan BCC, Gerlagh R, Klaassen G, Schrattenholzer L (2002) Endogenous technological change in climate change modelling. Energy Economics 24: 1-19

Welsch H (1996) Klimaschutz, Energiepolitik und Gesamtwirtschaft. Eine allgemeine Gleichgewichtsanalyse für die Europäische Union. München

Weyant JP (1993) Costs of Reducing Global Carbon Emissions. Journal of Economic Perspectives 7: 27-46

Zundel S., Erdmann G., Nill J., Sartorius C. and Weiner D. (2003) Innovation, Zeit und Nachhaltigkeit - Zeitstrategien ökologischer Innovationspolitik – der Forschungsansatz. In Horbach, J., Huber, J., Schulz, T. (eds.): Nachhaltigkeit und Innovation. Rahmenbedingungen für Umweltinnovationen. München

Zundel S., Erdmann G., Kemp R., Nill J. and Sartorius C. (2004) A conceptual framework for Ecological Innovation Policy. In Zundel, S., Sartorius, C. (eds.): Time Strategies for Innovation Policy towards Sustainability. Cheltenham

Appendix: The model structure of PANTA RHEI

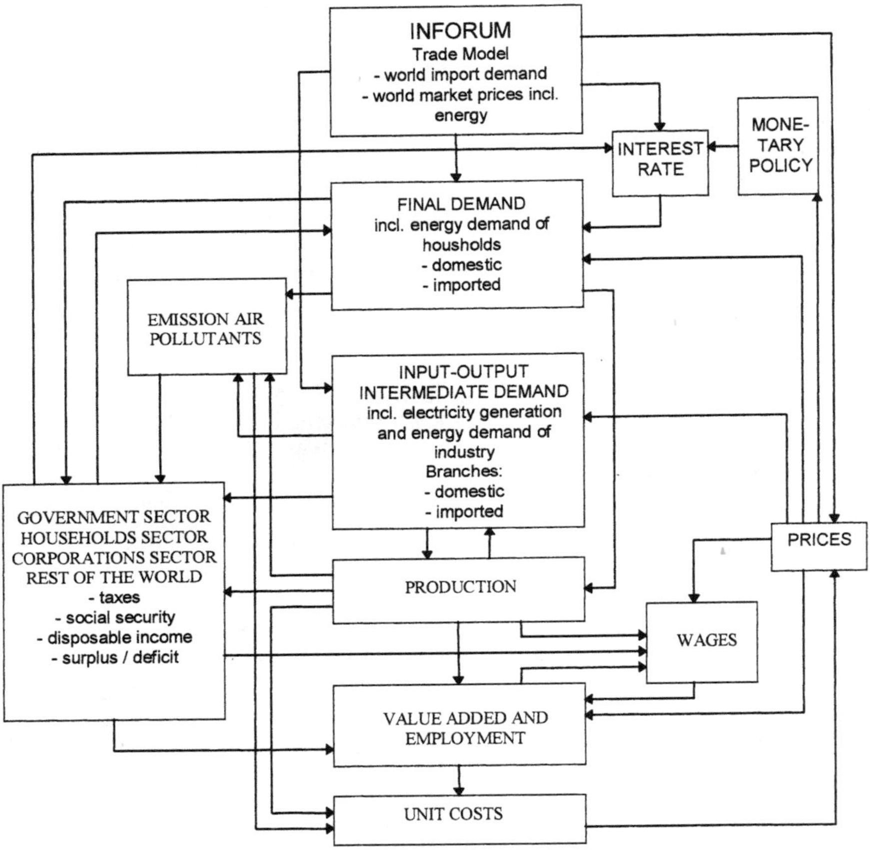

Indicators for Innovations Towards Prevention of Toxic Risks

Andreas Ahrens and Arnim von Gleich

1 Background

Reduction or elimination of long term toxic risks from manufacturing processes or the use of final products is one of the challenges for achieving more sustainable production and consumption. Toxic and eco-toxic risks have been publicly discussed for about 30 years, however the current approaches to the design of products and processes as well as the strategies to manage (eco-)toxic risks have not yet sufficiently mobilised market forces to eliminate exposure to hazardous chemicals. There are quite a number of success stories, but there are as well many cases where risk management has failed so far. For instance, this is the case with regard to i) the occurrence of man-made substances in polar regions, in other remote areas or in human breast milk , ii) the exposure of workers to dangerous chemicals in the construction and building sector or iii) indoor pollution from furniture and construction materials.

The SubChem project[1] aims to identify the driving forces and the dynamics behind success and failure in preventing the exposure to hazardous chemicals. 12 substitution cases are analysed as special types of innovation processes which are more or less explicitly determined to protect human health and the environment. Based on these case studies three questions are to be answered: In which way can enterprises, authorities and the civil society successfully interact with each other to prevent toxic risks (ability

---

[1] SubChem stands for "Sustainable Substitution of Hazardous Chemicals" and is one of 10 projects within the public funding program "Frameworks for Innovation towards Sustainability :[riw]" of the German Federal Ministry of Education and Research (BMBF). This research effort on political strategies, innovation systems, industries, technologies or policy instruments aims at a better understanding of the essential and promoting factors facilitating innovative solutions towards sustainable economies. SubChem is carried out in co-operation between the University of Bremen, the University of Applied Science Hamburg, the Institute for Environmental Strategies (Ökopol GmbH, Hamburg) and Co-operation-Centre Hamburg (a unit within the Department of Science of the City of Hamburg). www.subchem.de

to innovate)? How can the prevention of toxic risks (risk management) be integrated into quality management systems at company level, supply chain level or sector level? How can managers get more confidence and clarity on the "right direction" in preventing toxic risks along the life cycle of products (direction of innovation)?

In order to derive some general conclusions from the case studies and to establish sufficiently well documented cause-effect links, a basic model of "innovation systems" for the production and use of chemicals has been drawn up (see section 3). It contains the key drivers (Input), the architecture of the innovation system, typical interactions of key actors (process pattern) and the innovation-output. Based on this model, those framework conditions, relationships among key players and options to influence the system can be highlighted, which are likely to promote or hinder innovation towards prevention of toxic risks.

Such a model may be useful for the development of public policy and new regulatory approaches. However, more knowledge on the factors promoting the ability of systems to innovate may also contribute to a better understanding among companies of their own role in innovation systems and to identify new options to act that are not yet sufficiently explored. The more single actors will take a perspective on the "whole innovation system" the more it may be possible to undertake common steps towards increased capacity to innovate.

Targeting research to single companies or value chains (micro or meso level) is usually of an explorative nature and mainly limited to case studies. It is possible to identify external and internal determinants, to investigate interlinkages and to postulate cause-effect-relationships. However, due to the limited number of cases, it is difficult to derive general conclusions that are sufficiently representative for whole groups of companies or industrial sectors.

Based on the results of the case studies, it is a major aim of this paper to derive measurable indicators that can be validated by econometric models.

2 The future EU chemicals policy

In spring 2001, the EU Commission published the White Paper on the strategy for the future chemicals policy in Europe. The strategy aims to ensure a high level of protection of man and the environment against risks from the production and use of chemicals. At the same time the new policy shall provide a framework to increase competitiveness and innovation in the manufacture and use of chemicals. The policy has become necessary because of some serious shortcomings in the current system.

- For more than 30 years state institutions have tried to reduce the risks to human health (consumers and workers) and the environment related to the use of hazardous chemicals. Hazardous chemicals are often subject to several, overlapping regulatory regimes during their life cycle at both levels, production sites and products. This has generated a patchwork of various, partly inconsistent regulations. The implementation needs supervision in hundred thousands of companies.
- The EU existing substance regulation[2] has failed so far. Due to the absence of a phase-in regime with deadlines, new substances still have to fulfil much stricter testing requirements than substances which were placed on the market before 1981. The chemical industry regards the information requirements on new chemicals as a major obstacle to innovation in designing new industrial substances[3]. Even the assessment of a few priority substances at EU level is extremely slow since the burdens of the assessment process are on the authorities and there are no obligations for the users of these chemicals to contribute information for the assessment of risks. In addition to that, there is a lack of methodology for risk assessment and risk communication simple enough to be applied by the companies along the value chains
- The high profile of (eco)toxicity in the assessment of chemicals and the loss of trust in chemicals in the public has triggered high interaction-costs within trade and industry. This includes losses in the stock markets in case of public „toxic scandals“ or liability claims.

The chemicals policy reform will have a relevant impact on the whole manufacturing sector[4]. However, it makes a significant difference whether the stakeholders and policy makers consider the overall objectives mentioned above as conflicting goals (chemicals safety versus competitiveness) or as goals complementing each other. In other words: Will chemicals safety become an integral element in product and process quality

---

2 „Existing“ substances are substances which the chemical industry reported as existing in the EU market in 1981. These substances were regarded registered without further information requirements.

3 By definition, „industrial substances“ are all chemicals in processes and products except for pesticides, biocides, pharmaceuticals, cosmetics, food and feeding stuff.

4 In Germany, up to 20% of the production processes in industry (expressed as value added) may be affected by the reform; see ADL (2002)

(holistic approach) or will it remain a goal which can only be achieved by tough regulatory interventions?

The new framework for the production and use of (hazardous) substances may lead to an institutional innovation serving both, to achieve a level of protection for human health and the environment, and to promote the ability to innovate. However, this depends very much on whether the political debate results in a modern, regulatory system or whether the old mistakes are simply repeated, for example: Command and control by authorities instead of better access to risk related information and understanding in the markets and in the civil society as a whole. Or the other way round, excessive trust in market mechanisms instead of more efficient regulatory regimes.

3 How to predict or measure success?

If the overall goals are set, what are the indicators to measure success of the EU policy reform or national chemicals policies? The answers provided below are based on the 12 SubChem case studies and a brief review of the existing proposals for „chemicals indicators“ in Europe.

First of all, some key terms in our concept should be explained. Prevention of toxic risks and damage means reducing the probability of adverse impacts on human health and ecosystems due to exposure to (eco)toxic substances released from processes or products (losses, emission and discharges). Such adverse impacts can be expressed for example in i) "unit risks" related to exposure to carcinogens, ii) the rate of occupational diseases, rate of fish diseases or reproductive disturbance in wild-life, iii) the occurrence of contaminants in breast milk or iv) compensation costs related to occupational diseases, liability claims, cleaning costs for polluted sites and equipment.

In this context sustainable innovation leads to a decrease of overall risks (not shifting to other type of adverse impacts) and at the same time to increasing market shares and profit for companies offering safer solutions. This kind of innovation usually does not take place in single companies or as a response to isolated regulatory interventions but develops an innovation system responding to various external and internal forces (see figure 1).

An indicator is the measurable aspect of an input to such innovation systems (determinants) or the corresponding output (innovations and their impacts). Indicators should be analytically sound, easy to understand and should be based on available data. These three prerequisites in defining useful indicators considerably limit the number of options in particular with respect to the current data availability.

4 Determinants in the innovation system model

In setting up a basic model of "innovation systems" for the production and use of chemicals, a number of principal stakeholders in the supply chains can be defined: (1) the producers of substances, (2) the manufacturers of preparations (like paints, lubricants, dye stuffs, plastic compounds) by mixing single substances (formulators), (3) the industrial users of substances and preparations for manufacture of articles (like cars, furniture, electronic devices or textiles), (4) the professional users of preparations for e.g. cleaning or renovation of buildings and finally (5) the consumer. At each level of the supply chain chemicals traders (6) or retailers of articles (7) are involved. And finally at each stage of the life cycle of a product, emissions to water and waste have to be handled in the waste and waste water infrastructure (8). In addition, the manufacturers of the processing equipment (9), like for example metal cutting, printing or textile processing machinery often play an important role in the innovation systems for chemicals.

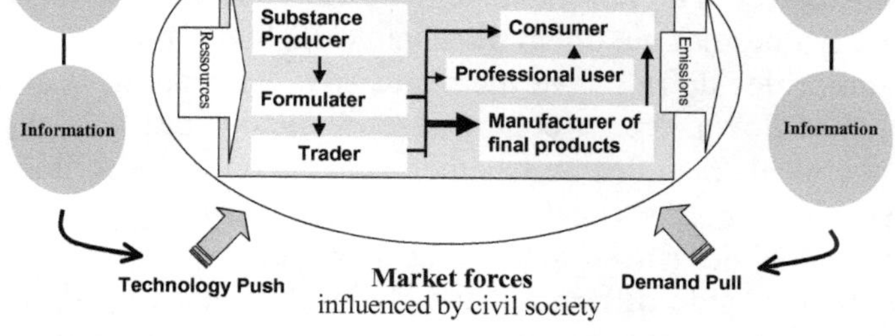

Fig. 1. Model of innovation systems

4.1 Drivers

In an ideal world, the interaction among the players in the supply chain may be mainly driven by technology push or by product/service demand. Both the supply and the demand side are directly influenced by the public debate on risks related to chemicals. Scandalisation of products or companies can seriously influence the demand side, in a broader sense: loss of market shares, loss of reputation, losses at stock markets.

At the same time, public policy, usually influenced by various stakeholders of the civil society, sets the regulatory framework. The supply side is regulated by the fully harmonised EU legislation on chemicals (classification and labelling, marketing and use restriction, risk assessment methodology, registration of chemicals). The demand side is regulated by EU minimum standard legislation on workers health protection and environmental protection.

However, public policy uses more instruments than direct regulatory restrictions. Goal setting and informational support programs as well as indirect regulatory tools: Making risk related information available in the market, shifting responsibility (duty of care) and costs from public administration into the market, setting up framework requirements on products, which are to be implemented as norms by private standardisation bodies.

4.2 Responsiveness of the system architecture

The readiness of a certain innovation system to respond to these drivers largely depends on the architecture of the system and the quality of interactions within the system (capability to innovate). If dangerous chemicals for example are used by many small professional users and additionally traders play a key role in disseminating information, usually the regulatory pull has little effect: The risk management advise needed to ensure compliance usually does not get down the chain (regulatory push). At the same time, the authorities do not have the capacity to enforce the requirements in thousands of down stream companies (regulatory pull). In the future, traders of chemicals may decide whether to make profit by selling tonnage under intransparent market conditions, or to make profit by providing good information and advise (including selling chemicals).

The situation is quite different if the chains are relatively short, the industrial users have a strong position in the market, the product has a high public profile and is for example under pressure from waste legislation (e.g. cars or electronic equipment). In such cases the industrial users of chemicals will actively search for innovative solutions and hence form an emerging market for the producers of chemicals.

Nevertheless, most players in the supply chains and industry sectors still have some difficulties to respond to policy instruments which require a certain level of co-operation and openness towards suppliers, customers and competitors. Innovation in the chemicals sector does not only depend on the inputs but also on the responsitivity of the innovation system. If enterprises do not set up their "radar systems" for emerging and differentiating demands of their customers they may fail to realise the opportunities for innovation.

5 The process

Each innovation process has its typical dynamics which also determine the overall output of the process. As an example, figure 2 gives an overview of the substitution process related to chlorinated solvents in Germany in the last 20 years.

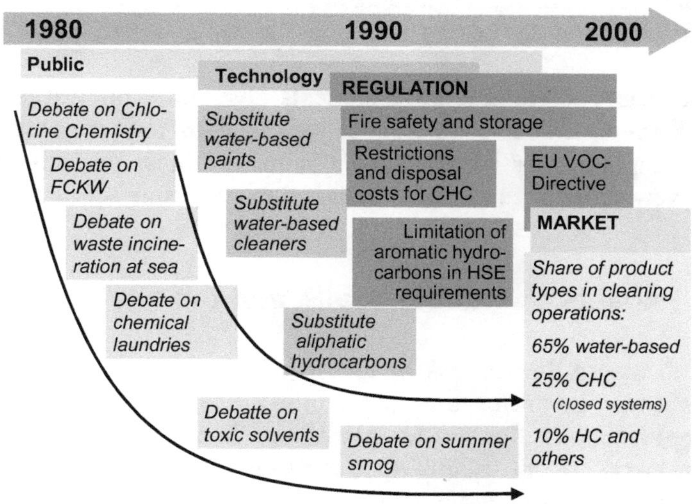

Fig. 2. Substitution process related to chlorinated solvents in Germany

Triggered by the public debate on chlorine chemistry in the 80s and the solvent issue as such, water based paints and water based cleaning systems for metal surfaces were developed. At the same time, the supply of chlorinated solvents for open use was transformed into another type of business: a chemical service based on a closed application system. In the 90s a tough regulatory regime pulled the market from 80% cleaning operations based on chlorinated solvents down to a share of 25% in 1996. The market volume e.g. of Tri and Per decreased by 90% during that period, indicating parallel reduction of emissions in the remaining applications. At the same time the share of water based cleaning increased up to 65%.

6 The output

The example of cleaning metallic surfaces shows that innovation may occur at different levels. In this case the innovation was not based on new chemical substances but on a new application system (including adapted machinery) for substances (surfactants and additives) which had been already in the market for a longer time. Parallel to this innovation process another important step were made with regard to solvent based cleaners: Selling the "function" has replaced selling the solvent itself. This leads to more efficiency and decreasing emissions (losses as waste, waste water or air emission).

In other cases like e.g. photo-chemicals, dye stuffs for textile, process regulators for chemical synthesis, adhesives with low concentration of volatile compounds or biodegradable mineral fibres, the innovation is based on new substances and new chemical mechanisms.

7 The impacts on human health and the environment

In order to assess the success or failure of relevant policies or research programs it may be useful to determine the outcome, the "quality" (width of the step) and "direction" of an innovation. Due to the chronic nature of many (eco)toxic effects, usually there is a time delay between the diffusion of a risk reducing innovation and their measurable impacts related to health and environment. Hence, risk assessment techniques are needed to predict the impacts of new products or new processes rather than waiting for the adverse impacts becoming measurable. There are different approaches to do this. For example: The potential impacts of a specific innovation can be evaluated against the concept of "inherent product (or service) safety", which is based on four essential product qualities:

- The chemicals should fulfil their technical functions in the most efficient way, hence minimal losses to waste, waste water and air should occur. For example, coating systems applied through spray application may still have an efficiency of less than 30% compared to powder coating where an application efficiency of nearly 100% can be achieved.
- The chemical product, the article or the service is either designed in a way that it can be easily absorbed in the metabolism of the ecosystems (e.g. like biodegradable loss lubricants for saw chains), or it is designed to be handled in "contained" systems throughout the whole life cycle (e.g. chlorinated solvents in the safechem-system). The "contained" system requirements include that the chemicals can be recycled back into the supply chains or disposed off in standard treatment facilities (waste water treatment, sewage sludge treatment or incinerators including slag treatment)
- The mobility of dangerous substances in a product or process is reduced to the technical possible minimum in order to prevent workers, consumers and the environment from being exposed. For example, this can be achieved by using pellets instead of powder or by using substances with a low vapour pressure or by reducing the tendency of a substance to migrate from matrices into indoor air or the environment.
- Products intended for wide disperse use in high volumes must not contain hazardous substances, in particular those for which a safe dose is very low or cannot be determined at all (e.g. carcinogens, mutagens, potent sensitisers, persistent and bioaccumulating substances). The risks related to such substances cannot be sufficiently controlled any more once they have been widely dispersed in technosphere and ecosphere.

To increase the inherent chemical safety of products not only toxic risks need to be taken into account but also changes in energy and water consumption, for instance.

8 Types of indicators

A study recently carried out by AEA Technology Environment for the DEFRA (Adams and Brown 2003) in the UK gives a comprehensive overview on "chemicals" indicators currently discussed and/or used at the European level. These indicators however are mostly based on the traditional OECD DPSR-Framework[5] and do not address innovation issues. Nevertheless some of the indicators can be used to measure certain inputs and outputs defined in the model above. The indicators can be clustered according to the following headlines[6], and combined with SubChem's indicator approach as illustrated in figure 3:

- Production (tonnage) of chemicals (of potential concern or substances under international phase out agreements) [indicator point 1]
- Release of (priority) substances (tonnage) into the environment (including those under international emission reduction agreement [indicator point 4]
- Generation and disposal of hazardous waste (tonnage) [indicator point 4]
- Exposure of wildlife and humans to hazardous substances (concentration in environmental media, biota, food and human breast milk) [indicator point 5]
- Impacts: selected biological effects in wildlife, humans and ecosystems as for example occupational diseases, allergies or breeding success of marine birds and reduction of fish diseases [indicator point 6]
- Responses and/or uses: changes in market volumes of hazardous substances due to reduction of losses (better containment) and/or substitution by less hazardous substances (including substances with obligatory reduction or elimination targets in industrial and consumer products); availability of sufficient risk information; [indicator point 2 and 3].

Most of the "chemicals" indicators currently used at the European level address the emissions, exposure and effects of chemicals or track implementation of phase out agreements for a few substances. The current indicator systems are not yet suitable to measure progress in preventing risks. Also, indicators to measure the key drivers or the readiness (responsiveness) of a system to innovate are largely missing in the current chemicals debate. This is partly due to a lack of clearly established (analytical sound) cause-effect links and partly due to a lack of interdisciplinary research in particular between environmental, chemical, political and economic research.

---

5 Driving Force, Pressure, State, Response Framework

6 Adams, M.L. (2003), page 13

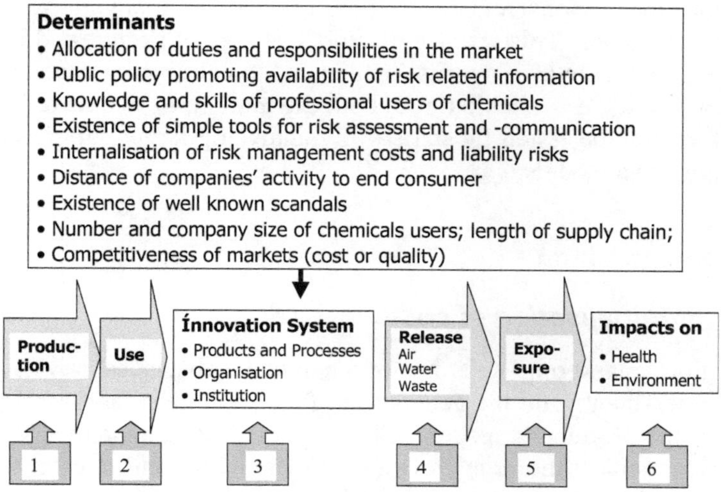

Fig. 3. Indicators related to health and environment

9 Options for innovation-indicators

There are enough indicators existing to measure the impacts of innovation or non-innovation at point 4-6 and 1-2. Although the availability of information on production and use of chemicals in the European market (1 and 2) could be largely improved, in principal the data is existing and could be used in an analytical sound way. This also applies to the use patterns of substances for which data exist at the level of producers, formulators and traders but has not yet been used to produce useful information. Most companies for example have not yet linked their management systems for i) purchase data, ii) material flows data and iii) data from the inventories of dangerous chemicals.

The situation is a bit different with regard to emission and exposure data. Although monitoring programs have been carried out all over Europe for more than three decades now, harmonisation of methodology is still a big issue. In particular the systematic selection and prioritisation of substances for measurements based on common criteria has only started in recent years. Much work is going on in this area, however, the basic indicators do exist and are being used.

Thus the following proposal will focus on a few indicators to measure the input, the responsiveness and the output of an innovation system related to chemicals. This set of indicators has not yet been tested in a quantitative model. However the case studies led to convincing conclusions and the data should be available throughout the market. In two next steps these indicators may be tested by statistically analysing larger samples and also in econometric models.

9.1 Determinants

9.1.1 Internalisation of costs

Production, marketing and use of chemical substances should include safety assessments and the generation of appropriate safety information as a part of the producer's responsibility and quality management. However, up to now neither the regulatory system (including enforcement) nor the voluntary schemes (product stewardship) are suitable frameworks to allocate responsibilities and costs in a way that toxic risk prevention pays off in a broad scale: Authorities carry the burdens of risk assessment for most substances in the market and the users of chemicals or the public carry the cost of occupational diseases, sick building syndrome, hazardous waste management, public waste disposal systems, waste water treatment, decontamination of buildings and polluted sites.

A regulatory framework shifting the costs of data collection, risk assessments and risk management (including insurance related to assessment and management mistakes) into the market may create an environment where i) good risk information or ii) elimination of hazardous substances or high-exposure-uses lead to down stream cost savings. Such a regulatory environment may also promote innovation towards risk prevention.

The most suitable indicators for such a process would be i) the share in companies' insurance fees directly or indirectly related to chemicals risks and ii) and to which extent chemicals users are ready to pay an extra price for good quality risk information and risk management advice.

9.1.2 Availability of good quality risk information

Every industrial or other professional user of chemicals makes choices among the products available in the market, usually based on price and technical performance. Whereas this information is usually readily available, information related to toxic risks is often of insufficient quality or missing at all. This is not only due to the lack of key data on the substances' properties but also due to the fact that many companies do not have sufficient knowledge and management capacity to produce (manufacturers of chemical products) or understand (users of chemicals) a standard safety data sheet, yet.

Systematic evaluation of data availability, documentation of progress in the current EU risk assessment program and safety data sheet quality are well established exercises in the EU. Statistics on information availability could be used as indicator for both, i) measuring success of the EU policy reform and ii) availability of one of the key prerequisites for innovation related to the prevention of toxic risks.

9.1.3 Risk information standard

The availability and understandability of risk related product information are prerequisites for making informed choices at the level of the supply chains. This includes the existence of appropriate and harmonised assessment and communication tools in the European (global) market. Based on that new product qualities can be developed and communicated in the market.

The extent to which public policy and sector policies promote the development and diffusion of such a standard in the market seems to determine widely the capacity of a system to innovate. The existence of such standards, the need to apply the standards and the actual coverage in the market can be measured by regulatory analysis, interviews or internet surveys.

9.1.4 Awareness pattern in the supply chain

Assumed that current or anticipated demands are key drivers for innovation, the distance of a company from consumers (or sensitive consumer groups), from well educated groups of workers, or from sensitive environments (e.g. touristic interest) should determine the profile of toxic issues in the companies management system. However, the awareness of the customers or the activity of public interest groups play a major role. Consumers for example have a hierarchy of toxic concerns with a focus on food and cosmetics. The quality management efforts in the manufacturing industries are accordingly high.

Compared to these fields, indoor building products are not yet an issue and hence prevention of indoor pollution is not yet regarded as a broadly recognised quality challenge in the building and construction sector. A large percentage of the professional users of chemicals far down the supply chain are not well educated to evaluate the risks and to handle chemicals in a safe way. The awareness in this area is not very high, too. As a consequence the suppliers in this sector are not challenged by an increasing demand for more product safety and better product information.

Also, in manufacturing sectors where the trade unions or the health insurance bodies play an active role related to the prevention of chemicals risks, the awareness among users of chemicals can be higher.

The awareness pattern in supply chains can be used as an indicator for potentially emerging markets. The number of relevant articles in special interest magazines, media coverage for smaller and larger scandals, the existence of public or private bodies systematically informing about dangerous substances in products or the existence of statistics on externalised costs can be used as indicators for innovation drivers.

9.1.5 Type of market

Markets of high volume, multi-purpose substances or preparations (e.g. cement, thinners, paint strippers, plasticisers in soft PVC-System) with many (often rather small customers) are usually driven by price competition and a well established system of industrial quality standards. Any additional costs for non-binding health, safety and environment measures are usually avoided. The processing machinery or work organisation is often well adapted to cost-efficiency but rather inflexible to changes in product design. In such markets innovation usually needs a strong – mainly regulatory-push, since neither sufficient enforcement on the users side (regulatory pull) nor an intrinsic readiness to pay a higher price for safer products are very likely.

Compared to that, the markets of performance chemicals and specialities with a limited number of customers (e.g. dye stuffs) are more open to competition on quality. This may include environment and health issues, in particular if the chemical using company sells consumer products (see awareness). In such markets environment and health related qualities may emerge also without strong regulatory measures, in particular if there is a strong global competition (e.g. textiles).

9.2 Innovation - output

The innovations related to the reduction of toxic risks and the consumption of chemicals in quantitative terms may be measured by the following types of indicators.

9.2.1 Selling the 'function' replaces selling of chemicals

There are several examples of "chemical services" having replaced the marketing of chemicals as such, like e.g. spray painting in the car industry, solvent based cleaning of metals or textiles or systematic hygienic consultancy in food and metal processing. With more and more quality diversity in chemicals product (specialised chemicals instead of multipurpose products) this component in the market is expected to increase. In such innovations the efficiency of product use increases and the risk of exposure due to losses naturally decreases at the same time.

Usually the information-component and the material-component in a product price (or service) are not easily to differentiate from each other. However, it should be possible to separate the raw material costs, the labour costs in the production and the labour costs in sales/marketing from each other. The product specific costs for generating good quality risk management information may be a bit more difficult to obtain at company level. Sector-specific innovation-output indicators could be based on the shift from production costs to information and consulting costs.

9.2.2 Product properties

Indicators for product quality related to environmental impacts are well established for eco-balancing based on life cycle assessment. A broad range of assessment methodologies are being applied by the manufacturing industry, in the framework of eco-labelling schemes and also as a basis for regulatory initiatives in the waste arena. However, the methodology of evaluating the risk of long-term toxic impacts along the life-cycle of a substance or a chemical product is not yet well established. Based on the critical impacts on health and environment, the following indicators may be used to measure innovation at the level of products:

- The losses of dangerous substances into the working environment, the indoor environment or the natural environment and hence the decreasing exposure potential (reduction of mobility and better containment). The Emission Scenario Documents as elaborated at EU and OECD level would be a good basis to determine the point in the life cycle to measure innovation.
- The technical performance of a product or service should be delivered based on the smallest possible impacts to health and environment. This applies to the amount of water, energy, raw material and chemicals, efficiency gains by recycling included. Various methodologies are available to measure innovation in this context.

- Substances used in systems open to the water environment should be sufficiently biodegradable. Based on the statistics for new (notified) chemical substances (a combination of substance properties and use patterns) it should be possible to follow up progress here. The same will be possible, when registration of existing substances takes place under the REACH requirements.
- Substances of very high concern (criteria are mostly agreed at EU level) should be totally eliminated from products in wide disperse use. Elimination of such substances can be measured at the level of producers or formulators (e.g. by interview or questionnaire) or at the level of users based on Safety Data Sheet information.

10 Summary and conclusions

In May 2001, the EU Commission published the White Paper on the strategy for the future chemicals policy in Europe. In October 2003 the Commission made corresponding legislative proposal. Hence, the overall goals are set, but what are the indicators to measure success of the EU policy reform or national chemicals policies? The answers provided in this paper are based on the 12 case studies carried out within the SubChem project and a brief review of the existing proposals for „chemicals indicators“ in Europe.

In order to derive some general conclusions from the case studies and to establish sufficiently well documented cause-effect links, a basic model of "innovation systems" for the production and use of chemicals has been drawn up. It contains the key drivers (Input), the architecture of the innovation system, typical interactions of key actors (process pattern) and the innovation-output. Based on this, the attempt was made to derive measurable indicators which may also help to validate the conclusions derived from the case studies in econometric models.

Most of the “chemicals” indicators currently used at the European level address the emissions, exposure and effects of chemicals or track implementation of phase out of a few substances. The current indicator systems are not yet suitable to measure progress in preventing risks. Also, indicators to measure the key drivers or the readiness (responsiveness) of a system to innovate are largely missing in the current chemicals debate. This is partly due to a lack of clearly established (analytical sound) cause-effect links and partly due to a lack of interdisciplinary research in particular between environmental, chemical, political and economic research.

The proposals in this paper focus on a few indicators to measure the input, the responsiveness and the output of an innovation system related to chemicals.

Four measurable determinants as "drivers" of the system have been identified: The degree of internalisation of costs related to risk management or related to the compensation of adverse effects, the availability of good quality risk information in the market, the awareness pattern specific for certain value chains and finally the specific market conditions (e.g. competition on quality or competition of price).

Two indicators were determined to measure the innovation output related to the reduction of toxic risks and the consumption of chemicals in quantitative terms: This is i) extent to which the selling of chemicals is replaced by the selling of functions and ii) the shift of product qualities in the market. For example, this may be the shift towards products with reduced losses of chemicals during their service life or the development of products with a better eco-performance.

This set of indicators has not yet been tested within a quantitative model. However, the case studies led to convincing conclusions and the data should be available throughout the market. In the next steps these indicators may be tested by statistical analyses of larger samples and by econometric methods.

11 References

Adams ML, Brown AK (2003) Performance Indicators for the UK Chemicals Strategy. Final Report (AEAT/ENV/R/1323) for the Department for Environment, Food and Rural Affairs. AEA Technology Environment (Oxfordshire); www.defra.gov.uk.environment/chemicals/csf/

Athur D.Little Ltd. (2002): Economic Effects of the EU Substances Policy; report on the BDI Research Project; Wiesbaden

Environmental Innovations in the Chemical Industry - Case Studies in a Historical Perspective

Melanie Monßen

1 Introduction

Industrial innovation processes are shaped and influenced by a multitude of factors. Usually the market and competition conditions, specific strategies of corporations, the mandatory law, and the political and societal environment have to be taken into consideration. Thus, the direction of the innovations and the motivation to undertake environmental innovation is considerably complex. Here the state takes part in a multifaceted way, for example by technological policy, environmental policy or the provision of public infrastructure. In this process public measures have to take the particularities of the specific innovation systems into account in order to influence an innovation into the direction of sustainable development. The research project „Co-operative institutions for a sustainable paradigm shift within the industrial sector - The example of the chemical industry" (COIN)[1], aims to survey internal sense-making processes that generate environmental innovations. Furthermore it considers the complexity of factors influencing innovation and environmental decisions in firms on the basis of operational case studies. The main interest focuses on the identification of influencial factors which push the development and implementation of environmental innovations.

Recently, there has been growing interest in researching approaches to economic, environmental and social performance measurement. Researchers and institutions seek to develop quantitative indicators for the sustainable economy as a whole. However, this idea gained substantial importance in the early 1990s as major institutions such as the Organisation for Economic Co-Operation and Development (OECD) and the European Commission began the process of defining innovation indicators and coordinating their implementation across countries. These initiatives led, for example, to the "OECD Work Programme on Sustainable Development".

1 Funded by the German Federal Ministry of Education and Research (BMBF) under the research initiative on sustainability and innovation.

This programme was aimed at the development of indicators for sustainable development and attempted to provide theoretical and methodological foundations and guidelines for new innovation indicators.[2] Another result of the initiatives was the Community Innovation Survey that was implemented by the European Commission via EUROSTAT in 1992-93 and revised in 1997-98.

Our research on an indicator system for environmental innovation should be seen in the context of this broader process of indicator development (Kemp et al. 2001).

This paper pursues a micro-perspective on this subject. The article concentrates on the description of determinants for environmental innovations in the chemical industry on the basis of source researched historical case studies. In a second step it identifies measurable indicators for environmental innovations.[3] The operational perspective of case studies – all of them researched at the Bayer Corporation in Leverkusen – offers specific examples of definite indicators for company related impacts of environmental innovations. The case studies provide indicators for process innovations (end-of-pipe-innovations as well as process integrated measures), product innovations and organisational innovations. As a matter of fact these case studies provide valuable insights in the innovation strategies of Bayer, even if it has to be considered that, characteristically, case studies may encounter difficulties in generalisation - arising from the specific, corporate point of view.

This article has the following structure: after a short overview on a selection of sustainable innovation indicator systems ('OECD Pressure-State-Response' model and 'Driving force-State-Response' model) in section 1, section 2 gives some information on the methodology of the COIN project. Section 3 provides three case studies which are summarised with respect to the derived indicator framework. The final section draws some conclusions. As the research project COIN is not yet completed this article can only provide preliminarily findings on the subject.

---

2 For more information on the OECD approach see Zieschank (2001) and OECD (1998).

3 These case studies resulted from the research project „Co-operative institutions for a sustainable paradigm shift within the chemical industry – The example of the chemical industry", funded by the German Federal Ministry of Education and Research (BMBF) under the research initiative on sustainability and innovation.

2 Indicator systems

Agenda 21, which was agreed on at the 1992 Earth Summit in Rio de Janeiro and was central focus of the 2002 World Summit on Sustainable Development (WSSD), called on both countries at the national level and governmental and non-governmental organisations at the international level, to develop the concept of indicators for sustainable development in order to identify such indicators.

The OECD approach on environmental indicators then developed one of the first sets of indicators which helped to understand changes in sustainable development. The OECD core set of environmental indicators is a commonly agreed on set of indicators for international use. The purpose of these is to keep track of environmental progress, to use indicators to measure environmental performance and to help determine whether countries are on track to sustainable development. The OECD Pressure-State-Response (PSR) framework states that human activities exert pressures (such as pollution emissions) on the environment, which can induce changes in the state of the environment (such as changes in ambient pollutant levels, water flows). Society then responds to changes in pressures or state with environmental or economic policies and programs intended to prevent or reduce pressures or environmental damage. The response component of the PSR framework relates to the actions taken by society, that are designed to prevent negative environmental impacts, to correct existing damage, or to conserve natural resources. These responses may include regulatory action, research expenditure, public opinion and consumer preference or changes in management strategies. Responses should be designed to act on the pressures.

Although the PSR framework is now widely used, it is being developed constantly. One of the main problems has been trying to differentiate between pressure and state indicators, and the need to expand the framework to deal more specifically with the needs for describing sustainable development. A development of the Pressure-State-Response framework has been the "Driving Force-State-Response" (DSR) framework selected by the United Nations Commission on Sustainable Development.

2.1 The conceptual model of Driving Force-State-Response

The Driving Force-State-Response (DSR) framework of the Commission on Sustainable Development of the United Nations is based on the environmental impact assessment model developed by the OECD in 1996. The DSR model is a multi-dimensional indicator system. At first it focuses on the human activities that create pressures. Thus in the DSR framework the term pressure, used in the former Pressure-State-Response framework, has been replaced by that of driving force in order to accommodate the addition of social, economic and institutional indicators. Moreover, the use of the term 'driving force' allows for the fact that the impact on sustainable development may be both, positive and negative, as this is often the case for social, economic and institutional indicators.

To sum up, the DSR framework is actually a matrix that incorporates three types of indicators horizontally (driving force, state, response) and the different dimensions of sustainable development vertically, namely social, economic, environmental and institutional. The feedback mechanisms that arise from information linkages between pressures and responses, between the state and the pressures, and from the state to the response give us the opportunity to better understand the consequences of policy and technology interventions.

Therefore the Driving force-State-Response approach is now widely applied for indicator frameworks which intend to define indicators for the regional or even national development of the environment. The Organization for Economic Co-operation and Development (OECD) and the European Commission have extended the DSR approach to find indicators for sustainable energy development. This way the framework is better tuned to the energy sector and makes good use of energy related environmental models that have been developed.

In order to find indicators for environmental innovations, which show the underlying forces of innovations, our indicator system applies the DSR approach as well. As the DSR indicator framework is orientated towards the macro perspective of environmental development, this is also true for our indicator framework. It delivers a broad comprehension of the whole innovation system including determinants for innovations (driving forces), an explicit description of the innovation itself as well as the regional and national impacts of an innovation system.

However, as the COIN project is orientated towards the micro perspective, the author concentrates on an analysis of indicators from an operational perspective. This implies that certain indicators discussed do probably not fit the regional or national perspective of the DSR approach. Especially when ecological impacts of operational innovations, which can have an effect for example on an improved climate, are concerned, the operational perspective is limited. For that reason this paper primarily concentrates on the research of indicators of driving forces for environmental innovations within our indicator framework, as these are definitely illustratable in the micro perspective.

2.2 Evaluation of indicators

Problems may result from the evaluation of indicator values in multi-dimensional indicator systems like this one. As it is likely to occur, we will have to weigh different indicators against each other.[4] As our case studies will show, a serious problem for assessing the different impacts of sustainable innovations results from the fact, that the evaluation of different indicators is very difficult. For example economic and ecological impacts of innovations often have to be weighed against each other.[5] And it seems even more difficult in the case of different ecological impacts that have to be weighed up – such as in our case of disposal of dilute acid (reduction of pollutant emissions against the increase of energy consumption). Our case studies will deliver some specific examples for this evaluation dilemma.

3 Methodology / case studies

In order to reveal findings about basis conditions that can push the development of sustainable innovations in the chemical industry, case studies of innovational processes at the Bayer Corporation were analysed on intensive source research in a historical perspective.[6] The chosen time frame for this analysis covers a century, beginning in 1901. The case selection reflects diverse innovation processes in the chemical industry and illustrates existing influence constellations. For significant cases of product and process innovations as well as organisational renewals the specific technological, market and business framework on the one hand and the societal and political influence on the other hand were worked out. For example the effects of new guiding principles, public technological programs and a changing statutory framework on the genesis and diffusion of new innovation paths have been examined.

---

4 "Evaluation problems can be recognised during the selection of the indicators, in the interaction between the indicators and also when interpreting the results of the developments described by the indicators", see the paper of Jens Horbach in this book.

5 See the paper of Jens Horbach in this book.

6 Although the collection of data by surveys or case studies is regarded as a very costly or even elaborate method and therefore can be regarded as merely inadequate for the monitoring of sustainable innovations in the long run, it was seen as a promising way to gain data on determinants for significant environmental innovations in the chemical industry.

Bayer's internal as well as the external activities were analysed with adequate empirical methods of economics (Pohl 1999; Pierenkemper 2000; Borscheid 2001; Platthaus 1999; Jeske 2000). The inner logic of economic action, this means the processes of decision making and the question of economic success, is the focus of historical business research like this (Pierenkemper 1999). As business decisions are made within extremely complex fields of circumstances, the company's connections to social sub-systems such as the media, non-governmental organisations and the chemistry industry council (VCI) are considered as well (Pierenkemper 1995). Against this background the following case studies have been carefully selected on the basis of certain selection criteria[7]:

1. Foundation of a waste water commission at the Farbenfabriken vorm. Friedr. Bayer & Co[8] in 1901
2. Termination of dilute acid dumping in 1982 – basis conditions and Bayer's alternative solutions for dilute acid disposal
3. Product innovation of iminodisuccinate (IDS) as substitute for ethylene diamine tetraacidic acid (EDTA)

Based on the results of these three case studies (one further study on the product innovation of polyaspartic acid as a substitute for polyacrylate is already in progress) a number of indicators for environmental innovations could be defined for each of the three levels of analysis (determinants, innovation system, impacts). However, as for the main orientation of the research project COIN, there will be a slight focus on the identification of determinants of environmental innovations and the definition of indicators for these innovations. The range of case studies enables us to describe innovations in processes as well as product innovations and organisational innovations.

[7] For the selection of case studies the following criteria have been established: impact of the innovation on the relief of environmental damage, kind of innovation (organisational-, process- or product innovation), preliminary social/public pressure on the company as well as the time frame. Concerning the time frame the main focus is on the past 30 years, regarding the commencement of the modern environmental policy in Germany. Only one case study was chosen covering a time frame at the beginning of the past century in order to have the chance to show a development direction of the significance of environmental affairs in the chemical industry.

[8] Hereafter called Bayer Corporation. The company name Bayer corporation was determined at the resurrection of the company after the decomposition of the IG Farbenindustrie AG in 1949.

4 Findings from the case studies

4.1 Initiation of a waste water commission at the Bayer Corporation in 1901

Effective links between organisational and technological innovations are crucial to a successful development and application of many types of technologies (Kemp et al. 2001). This is also true for our case study of the initiation of a liquid waste commission at Bayer in 1901 as the following section shows.

The foundation of the Bayer waste water commission, an analysing committee of the company's liquid waste, is regarded as an organisational innovation because this measure had never occurred before, neither at the Bayer Corporation nor at any other company of the chemical industries. The task of this commission was to collect required data on the company's liquid waste and to answer questions on the quantity as well as on the quality (e.g. acidic concentration) of liquid waste that was completely disposed off into the Rhine. In a further step the committee members, all of them specialists in their field of science (chemistry, engineering, business), had to develop innovative concepts of waste water treatment that would cause less water pollution and fish kill (Bayer corporation BAL 58/9.4.1).

4.1.1 Innovation system

The innovations which were initiated by the committee concentrate on end-of-pipe process innovations. The waste water discharge had been technically modified: instead of formerly two just one liquid waste sewer was used. The sewer for alkaline waste was disabled and the alkaline waste was mixed up with the acidic waste just before the discharge into the Rhine. This led to a less acidic and therefore less pollutant mixing ratio. Additionally collecting tanks were installed and the sewers were reconstructed in order to achieve a more constant discharge into the Rhine (Bayer corporation BAL 58/9.4.1). Therefore it can be regarded as an environmental innovation: as described the foundation of waste water commission led to an improvement of end-of-pipe waste disposal measures and therefore to a significant and measurable ecological discharge – as we will examine later in this section. However, let us first have a look at the determinants causing the innovation.

4.1.2 Determinants

Changes in national regulation
The intervention of the regional council (Regionaldirektion) Düsseldorf caused Bayer to take measures concerning its liquid waste disposal. A ministerial court order (Ministeriatsbeschluss) of February, 21st 1901 adjudged the authority to the commissioners to assay Bayer's industrial liquid waste at regular intervals. Additionally they were, for the first time, authorised to apply sanctions against Bayer - in case of objections. They requested for the first time scientific data on the quality and the quantity of the liquid waste that was discharged into the Rhine. This order forced the company to react and was followed by the foundation of a commission.

Protest against Rhine pollution by fishermen / activity groups
The complaint of the fishermen who lived (and worked) at the adjoining premises near the company plays a decisive role in this context, too. On the one hand the complaints of fishermen (or the fishery protection associations) drew the regional council's attention to the pollution of the Rhine.[9] On the other hand Bayer settled payments to those fishermen who complained about the loss of their essential income caused by the decreasing fish population in the Rhine. The amount of these payments or the chance of a potential cancellation of these payments respectively, caused an additional need for the development of concepts of waste water treatment for economic reasons.

4.1.3 Impacts

Ecological Impacts
In our operational perspective of analysis the ecological impact of an environmental innovation is harder to detect than the economic impacts – even if the period under consideration is about a hundred years ago. Nevertheless we can stress the effect of a significant reduction of pollutant concentrations in liquid waste as an indicator for ecological impacts of an organisational innovation.

As a consequence of the findings of the commission / the commission's certificate, numerous measures were taken to achieve a better constancy in the acidic concentration of the liquid waste. This achievement of a more

---

9 Even if a proof for such payments, which were provided for those fishermen who complained about their loss of earnings, can actually be found since 1909, there are hints about such payments for the preceeding times (Bayer corporation BAL 58/9.4.1).

consistant acidic concentration supported the described self-purifying power of the Rhine.

Moreover, the quantity of the acidic waste could be reduced by a change of production processes[10]: this caused a concentration reduction of the acidic emissions to less than one third (22,7 per cent) in December 1903 compared to the acidic concentration two years before.[11]

Economic impacts
In this case, the economic impacts of the organisational innovation seem to be irrelevant for financial accounting purposes. The part-time appointment of three Bayer-employees (1 engineer, 1 analytical chemist, 1 merchant) to take part in the commission did not mean remarkable losses of resources for the company. Only the engagement of an expert, who enjoyed an established reputation (Curt Weigelt), caused further investment for the writing of the expertise. Regrettably the exact amount of investment can't be reconstructed by the available data as the time period of this study lies almost a hundred years in the past. Therefore no specific indicator can be delivered. However, it seems that the company's investments were only driven by the prospect that on the one hand the payments of compensation for the adjoining fishermen could be reduced in the medium term and that on the other hand Bayer could defy the control on part of the local authorities by the implementation of initial measures that reduced the objected high concentrations of pollutants in the liquid waste.

4.2 Termination of dilute acid dumping in 1982 - basis conditions and Bayer's alternative solutions for dilute acid disposal

4.2.1 Innovation system

Against the background of image campaigns initiated by non-governmental organisations the Bayer management board decided in May 1980 a gradual reduction of Bayer's demand for ocean dumping. Bayer

[10] „Die Herabminderung des früheren starken Säuregehaltes ist also zurückzuführen sowohl auf eine nicht unbeträchtliche Säureersparnis im Betrieb, wie auch eine stärkere Verdünnung, welche sich pro Tag ausdrückt durch eine Mehrförderung von 8711 cbm Wasser." Note of commission agent Curt Weigelt to Bayer (Düsseldorf government 35948 353/6b). According to conversions of production see note of Carl Duisberg directed to Curt Weigelt of May 19th, 1904 (Düsseldorf government 35948 353/6b).

[11] Note of commission agent Curt Weigelt to Bayer (Düsseldorf government 35948 353/6b).

aimed to achieve abdication of ocean dumping by 1984 – based on a need for dumping of 166.000 tons in spring 1980 (Grünewald 1980). This was planned to be realised by creating alternatives in disposal engineering such as the construction of a treatment plant for organic dilute acid (BAL 58/9.4.8; interview with Dr. Frank-Andreas Schendel April 16th, 2003). During the development of such a new treatment plant Bayer benefited from technical procedures for the discharge of allied acidic materials implemented in Bayer's site in Krefeld Uerdingen.[12] Eventually Bayer succeeded in solving its disposal problem by 1982.

The innovation system is regarded as an environmental innovation as the new developed plant had the capacity to recycle the accumulated organic dilute acid completely. As a consequence the discharge into the North Sea became obsolete for Bayer by March 20th 1982 – two years earlier than expected and seven years earlier than competitors like Sachtleben or Kronos could manage. They stopped ocean dumping by 1989 (BAL 58/9.4.6). An assumed damage to the North Sea caused by the dumping of about 150.000 tons/year could be avoided by this innovation. The question is: what determinants provoked Bayer into developing an alternative technique of liquid waste disposal?

4.2.2 Determinants

Activity of NGOs

Firstly, an increasing social awareness of environmental problems has to be mentioned as well as a correspondingly increasing media presence. Due to the extensive and complex logistics, necessary for the transport of dilute acid to Rotterdam, Bayer was dependent on external service providers and thus became vulnerable.[13] These circumstances were for the first time utilised by non-governmental organisations like Greenpeace[14] in May 1980. Greenpeace organised a protest campaign against ocean dumping at Rotterdam harbour. For three days campaign groups blocked the transport ship that had loaded Bayer's dilute acid (BAL 58/9.4.7). This was the starting point of an international protest campaign against ocean dumping. Despite the fact, that Bayer was able to sue for a preliminary injunction against the

---

[12] Such technical procedures for the reprocessing treatment of inorganic dilute acid (waste product of the production of titanium dioxide) were used in Krefeld since 1967.

[13] Since 1972 there was a co-operation with Lehnkering Corporation (Duisburg). Lehnkering was responsible for the shipment of the dilute acid to Rotterdam.

[14] A German subsidiary of Greenpeace was founded in Hamburg on January, 1st 1980.

campaign group, the corporate image was already discredited for media attention was drawn on Bayer as an actor of ocean dumping.

Beside Greenpeace as main actor of environmental protection campaigns in 1980, a considerable number of non governmental organisations with their main focus on discharge of dilute acid, protection of the North Sea or just the company Bayer were founded. Approximately 1000 action groups existed or were founded in 1980. They organised a number of campaigns, part of them directly aimed against Bayer.[15] The most important action groups to be mentioned are the "Arbeitskreis Chemische Industrie Köln", the action group for environmetal protection in Leverkusen (Bürgerinitiative gegen Umweltgefährdung in Leverkusen, „Rettet den Rhein", „Wuppertaler Bürgerinitiative gegen Bayer-Umweltgefährdung Wuppertal", „Arbeitskreis Umweltschutz Brunsbüttel" and last but not least the „Bundesverband Bürgerinitiativen Umweltschutz e.V. (BBU) Karlsruhe" the umbrella organisation of approximately 1000 action groups in Germany.

Due to these activities the topic of ocean dumping in the North Sea gained considerable media presence in the time to come. The increasing public pressure caused damage to Bayer's image and is therefore regarded as a main determinant for the development of environmental innovations. In this case media attention as well as the number of protest campaigns against the chemical industry can function as amplifier of social attitude towards the topic of ocean dumping.

Regulations

Secondly, changes in national regulation led Bayer to a more sensitive view on the question of ocean dumping. The German hydrographic institute (Deutsches Hydrographisches Institut Hamburg DHI), the responsible state department granting licences for liquid waste dumping into the North Sea, assessed constrictions to the licence in 1980 for the first time (Federal Agency for Shipping and Hydrographic 1990). The assessment was due to a concern, that acidic liquid waste discharged into the North Sea causes damage to marine fauna such as plaices, as findings of an investigation of the German research institute for fishery (Bundesforschungsanstalt für Fischerei) pointed out (Dethlefsen 1986; Federal Agency for Shipping and Hydrographic 1990).

---

[15] Descriptions and documents concerning the Greenpeace protest campaign against Bayer are documented at the archive (BAL 58/ 9.4.5).

Not only in Germany but also in the Netherlands the licensing procedure was constricted in 1980. The Dutch administration (Raad van Staate), responsible for the licensing of waste disposal, made it more difficult for Bayer to receive the needed licence for ocean dumping into the North Sea. This was relevant for Bayer because in this case the dumping area was set 20 kilometres from the Dutch coast (near Scheveningen). This assessment was also due to the concern, that acidic liquid waste discharged into the North Sea causes damage to marine fauna - favoured for example by the Dutch organisation "Natuur en Milieu" (nature and environment) that actually sued against the permission of ocean dumping in summer 1980 and reached an accentuation of the licensing process. As for these reasons further licences were only given under a number of conditions[16] and had to be requested again every two years in a complex procedure.

The changing legal framework in Germany and the Netherlands as well as the purposeful protest campaigns by NGOs and the following increasing media coverage were the decisive determinants for the development of environmental innovations like the new treatment plant for dilute acid in Leverkusen.

4.2.3 Impacts

The reduction of pollutant concentrations in the southern part of the North Sea (Dethlefsen 1987) - here increased by 166.000 t of Bayer dilute acid per year - is considered a most remarkable ecological impact in this case. The assumed damage to the North Sea caused by ocean dumping has been reduced by this innovation.

At the actual state of information converse ecological effects, e.g. an increasing energy consumption caused by the developed treatment plant, are not regarded to depreciate the described ecological impact (reduction of pollutant concentrations in the North Sea).

4.3 Development and market introduction of IDS

Iminodisuccinate (IDS) is a medium-strong complexing agent that can substitute EDTA in cases where medium-strong chelating agents can achieve the objective to mask disturbing metal ions (Jobst 2000). In technical applications, IDS is used in detergents and cleaners, in textile and paper production and in the photographic industry and agriculture. IDS

---

16 Such as e.g. detailed information on the progress in developing alternative disposal concepts (BAL 9.4.8).

was applied for patent (Neustoff) in 1994 and is being sold with the trade name Baypure CX®, since 2001. Bayer remains the only producer of IDS. It is being produced with relatively benign chemicals with no harmful by-products and is biodegradable (Mitschker 2000). IDS has no serious negative environmental impacts. Its toxicological and eco-toxic effects are minimal according to present tests. Regarding these properties IDS is superior to other EDTA substitutes such as NTA, the use of which is not recommended or even illicit in certain European countries such as in Italy and in Turkey because of its potential hazardous impacts.

4.3.1 Innovation system

Against the background of some R&D progress in the development of the polymer polyaspartic acid (PAA), an environmentally friendly dispersing agent, there was soon the idea of a testing production of the related monomer of PAA (Moritz 2003). The product attributes of this monomer were researched and an affinity of the monomer, named Iminodisuccinate, to the chemical structure of EDTA (and other complexing agents such as DTPA) was identified. This was in 1993. Since then the monomer IDS was systematically produced by a variation of the ingredients of polyaspartic acid, namely ammonia, sodium hydroxide and maleic acid anhydride (Reisch 2002). The related production of complexing agents seemed to offer a promising option because they have a variety of application areas (Moritz 2003; Jobst 2000).

The production plant for IDS was well developed and a lot of process innovations were realised concerning improved processing technologies, closed cycle arrangements and waste water treatment (Moritz 2003). The R&D costs amounted to around 25 million Euro between 1992 and 1997 for Bayer including personnel from research, technical engineering and marketing efforts (Jobst 2000).

4.3.2 Determinants

Economic demand to process a basic chemical on the basis of maleic anhydride

Since 1993 a lot of maleic acid anhydride, a basic chemical that is primarily used as softening agent, has been produced in two large plants in Baytown, USA (business division organic chemicals). In view of low market prices for basic chemicals like maleic acid anhydride it appeared reasonable for Bayer to process this basic chemical into marketable fine chemicals (Moritz 2003; Jobst 2000). For this reason some R&D work on possible downstream products of maleic acid anhydride was done in Bayer's site in Krefeld Uerdingen.

Market opportunities
Beside the internal economic incentive an increasing demand for EDTA substitutes as well as the increasing framework conditions in favour of waste minimisation played a decisive role in the question of determinants. A favourable analysis of market opportunities of alternative new complexing agents became decisive for the successful development of IDS. This is also true for the development of polyaspartic acid, a polymer of IDS (Reisch 2002). In this case no relevant environmental concern and no political pressure to reduce EDTA release affected Bayer, for Bayer is no producer of EDTA.

In view of the economic market demands it seems reasonable to conclude, that Bayer was only willing to make these considerable investments because it perceived a good opportunity for a new medium strong complexing agent on the global market. This kind of R&D investment could only be justified economically from a global market perspective, referring to OECD countries in particular (Jobst 2000). From this perspective the German EDTA debate was a major ingredient, but not of decisive importance for Bayer's corporate strategy (Jobst 2000).

4.3.3 Impacts

In many industrial and household applications complexing agents enter the environment with the waste water. Regarding the fact, that IDS meets strict requirements in terms of eco-toxicity and biodegradability and that the manufacture of the product does not generate any waste requiring disposal, it is superior to conventional complexing agents such as EDTA which is not readily biodegradable or NTA which has potential cancerogenic impacts (Jobst 2000). Indicators for the ecological impact of the use and production of IDS may be the development of EDTA concentrations in rivers – mainly near the production sites of EDTA – or the ground water as complexing agents are essentially used by end-consumers e.g. in washing powder or for cosmetic purposes.

5 Conclusions

The aim of this paper was to analyse driving forces for environmental innovations from an operational perspective. And as initially expected our specific case studies show that Bayer's motivation to undertake environmental innovations were primarily determined by external factors. The innovation processes researched were driven by environmental policy regulation (establishment of waste water commission / disposal of dilute acid), by competitive factors or market demands (IDS) or as a result of social awareness of the need for clean production (dilute acid). In all these cases environmental innovations were undertaken even if it seemed not to be profitable. In short-term, considerable investments had to be realised. At first glance, this contradicts the scientific assumption that in an operational perspective the incentive to innovate is based on the company's expectations of a higher profit level or - at a minimum - maintaining a satisfactory return on investment (Kemp et al. 2001). However, even if the investment incentive was initially driven by other - not primarily profit orientated - determinants, such as in our cases, in the end all these factors affect the company's aim of entrepreneurial prosperity in the long run. That is why the economic sustainability is assumed as precondition for the implementation of innovation systems in our operational point of view.

The described indicators for environmental innovations only partially fit into the indicator framework based on the DSR approach. Most of them are indicators for the corporate effects of the innovations, image effects or cost reductions. In this micro perspective e.g. the dispersion of certain environmental effects cannot be sufficiently analysed. Though this does not mean that the innovations of the Bayer Corporation did not have any macro economic or ecological effects at all, in micro perspective case studies these effects can only be described rudimentary.

6 References

Bayer Corporation (ed) (BAL 58/9.4.1) Initial Report of the Waste-Water-Commission, November 9th, 1901. Archive of the Bayer Corporation

Bayer Corporation (ed) (BAL 58/9.4.1) Final Report of the Waste-Water-Commission, October, 24th 1904. Archive of the Bayer Corporation

Bayer Corporation (ed) (BAL 58/9.4.6) Press Release of March, 20th, 1982. Archive of the Bayer Corporation

Bayer Corporation (ed) (BAL 58/9.4.7) Report of the Department AWALU. Archive of the Bayer Corporation

Bayer Corporation (ed) (BAL 58/9.4.8) Waste Management Concept concerning dilute acid. Archive of the Bayer Corporation
Bayer Corporation (ed) Press release of the Bayer Corporation on June, 18th 2002
Bayer Corporation (ed) Baypure - neuer Name für bewährte Performance Chemikalien. Press release of Bayer on December, 7th 2000
Boscheid P (2001) Der ökonomische Kern der Unternehmensgeschichte. Zeitschrift für Unternehmensgeschichte 46: 3-10
Conrad J (2000) Environmental Policy Regulation by Voluntary Agreements: Technical Innovations for Reducing Use and Emission of EDTA. FFU Report 00-04
Dethlefsen V (1986) Überblick über Auswirkungen der Verklappung von Abfällen aus der Titandioxidproduktion in der deutschen Bucht. Hamburg
Dethlefsen V (1987) Gift ahoi - Nordsee tot. Auch die zweite Nordseeschutzkonferenz wird das Sterben des Meeres nicht aufhalten. Ein Interview. Die Zeit: 48: 16
Düsseldorf Government (ed) (35948 353/6b) Note of Carl Duisberg directed to Curt Weigelt on May 19th, 1904. Hauptstaatsarchiv Düsseldorf
Düsseldorf Government (ed) (35948 353/6b) Note of Commission Agent Curt Weigelt to Bayer. Hauptstaatsarchiv Düsseldorf
Federal Agency for Shipping and Hydrographic ((BSH) Bundesamt für Seeschifffahrt und Hydrographie) (1990) Abfallbeseitigung auf Hoher See. In: BSH (ed) Annual Report. Hamburg
Grünewald H (1980) Das „Problem Dünnsäure" bald technisch lösbar. Statement at the shareholders' meeting on June, 19th. In: Bayer Corporation (ed) Archive of the Bayer Corporation: BAL 58/9.4.5
Henneking R (1998) Chemische Industrie und Umwelt. Stuttgart
Horbach J (this volume) Methodological Aspects of an Indicator System for Sustainable Innovation
Jeske J (2000) Wozu Unternehmensgeschichte? Frankfurter Allgemeine Zeitung September 19th: 17
Kemp R, Arundel A, Smith K (2001) Survey indicators for environmental innovation. (Paper presented at the conference "Towards Environmental Innovation Systems" in Garmisch Partenkirchen)
OECD (1998) Environmental Indicators. Towards Sustainable Development. Paris
Pierenkemper T (1995) Gebunden an zwei Kulturen. Zum Standort der modernen Wirtschaftsgeschichte im Spektrum der Wissenschaften. In: Jahrbuch für Wirtschaftsgeschichte 1995/II. Köln
Pierenkemper T (1999) Was kann eine moderne Unternehmensgeschichte leisten? Und was sollte sie tunlichst vermeiden? Zeitschrift für Unternehmensgeschichte 44: 15-31
Pierenkemper T (2000) Sechs Thesen zum gegenwärtigen Stand der deutschen Unternehmensgeschichtsschreibung. Eine Entgegnung auf Manfred Pohl. Zeitschrift für Unternehmensgeschichte 45: pp. 158-166
Platthaus A (1999) Wo ist mein Platz? Die Unternehmensgeschichte soll neu begründet werden. Frankfurter Allgemeine Zeitung July 14th: 5

Pohl M (1999) Zwischen Weihrauch und Wissenschaft? Zum Standort der modernen Unternehmensgeschichte. Eine Replik auf Toni Pierenkemper. Zeitschrift für Unternehmensgeschichte 44: 150-163
Reisch MS (2002) Polyaspartic Acid. Chemical & Engineering News 80(8): 23-28
Zieschank R (2001) Zum aktuellen Stand von Umweltindikatoren im Kontext einer nachhaltigen Entwicklung. FFU Report 01-06. http://www.fu-berlin.de/ffu/Publikationen/index.htm

A Sustainable Innovation Scorecard for the Electronics Industry Innovation System

Joachim Hafkesbrink and Marianne Halstrick-Schwenk

1 Introduction

Sustainability which has been dominating environmental and development policy discussions for several years is considered as a key term for a future that combines economic progress with conservation of the environment and social justice. In order to take account of its ecological concerns both German waste management policy and EU environmental policy are following the paradigm of the circular flow economy, standardising extended producer responsibility for manufacturers or sellers of certain product groups. The most important instrument they draw upon in the assignment of this product responsibility is the take-back obligation, by which the manufacturer or seller of the relevant product is burdened with the costs of its disposal and the eco-efficient desirable combination of waste avoidance, recovery and disposal is to be achieved.

After having introduced take-back regulations for the category groups packaging, batteries and automobiles now for electrical and electronic equipment which present a great environmental problem due to the high amounts of waste to dispose, the range of the used materials and the resulting risks an analogical directive has been entered into force on 27 January 2003[1]. This has to be put into national law until September 2004. For Germany the Federal Ministry for Environment, Conservation and Reactor Safety (Bundesministerium für Umwelt, Naturschutz und Reaktorsicherheit 2003) has presented an outline of the general conditions for the application of the guidelines in Germany in April 2003.

However, the mechanisms triggering innovations for the achievement of environmental aims and determining their dimensions are relatively complex (Klemmer et al. 1999; Hafkesbrink et al. 1998). They are influenced by many different determinants and their mutual influence. Further legal regulations in the near future regarding waste policy will be the RoHS and the planned EuP, which will influence innovations for the improvement of waste avoidance, recovery and disposal of electrical and electronic equip-

---

[1] see European Parliament and Council (2003) EG ABl. L 37/24

ment as well. Beyond that other factors such as the availability of respective recycling technologies, the environmental awareness of consumers and companies as well as other market influenced impulses (e.g. demand of customers regarding the practicability of products) will determine the nature and direction of innovations.

An essential aspect of the :[riw]-project "Internalisation versus Internationalisation" (INVERSI)[2] is at first point out dimensions and direction of innovations which are started by the application of the WEEE-directive as important changes of institutional conditions in interaction with other relevant determinants and then evaluate these innovations with regard to their contribution to a sustainable development. INVERSI assumes a comprehensive understanding of innovations and also considers changes of institutional conditions as innovations.

A complex indicator system for the assessment of the innovation system regarding the context of efficiency of different framings of such an instrument is necessary. The concept of a Sustainable Innovation Scorecard (SISC) will be presented as one possibility for demonstrating such contextual relations. The (SISC) approach presented in this paper is a result of the study "ECOLIFE 2", a thematic network funded under the EU – Life Program, set up to co-ordinate national R&D projects in Europe in the electronics industry (Hafkesbrink to be published 2003) but reflects also a constant experience exchange between this project and INVERSI. This scorecard is a comprehensive (indicator) framework for the electrotechnical and electronic industry (electronics industry) on the impact chain of external drivers – behavioural change – sustainability impacts and can as well serve as a roadmap to develop hypothesises on different questions in this area. The scorecard can be divided into the three modules "indicators concerning determinants and their influence on innovations", "indicators concerning the presentation of innovations and their evaluation concerning sustainable development" and "indicators concerning the description of innovations according to their stage of development and diffusion barriers and derivation of "need of action".

For questions related to INVERSI mainly indicators from the first two modules are of relevance, as the focus is at the influence of the WEEE-directive on the achievement of innovations and the evaluation of their sustainability effects. Module 2 simultaneously allows a detailed considera-

---

2 INVERSI deals with the adaptation and arrangement of take-back duties considering the expected growing direct crossborder trading favoured by the use of new media. A growth is mainly expected for electrotechnical articles, this problem has already been taken into account in the WEEE-directive and the respective details will have to be considered in the implementation.

tion of possible innovations in the field of electronics. Module 3 provides the basis for a detailed analysis of innovations in order to derive necessary activities (need for action). Especially in this context empirical examples of ECOLIFE 2 are comprised. Before giving a detailed presentation of the concept with corresponding examples some general statements concerning sustainable development, sustainable innovations and product oriented waste management as well as the eco-political context will be given.

2 Sustainability, sustainable innovations and product-related waste management

2.1 Sustainability and sustainable innovations

During the last years sustainable development has become the key term, that denotes a future society and also combines economic progress with the preservation of the environment and with social equity. The concepts worked out in this context in general have a normative basis which consists of the demand for justice in and between generations. The starting point for concepts on intergenerational allocation justice were reflections on the handling of regenerative and non-regenerative resources as well as the absorption capacity of environment media. The management rules (Daly 1990; Pearce and Turner 1990; Enquete-Commission „Schutz des Menschen und der Umwelt“ 1993, 1994) developed then however, are only a first step of the realisation of this vision. Based on the fact that the future development of a society may be affected by ecological risks and economic distortions as well as social tensions during recent years the importance of the last two points has been emphasised more and more. The pragmatic concept of the so called three pillar model follows these ideas which include the principal equity of ecological, economical and social aspects (Klemmer 1994, 1999). The use of the three pillar model has significant advantages: the separate identification of ecological, social and economic objectives emphasises their independence and equal importance. Furthermore it is an open approach as the elements of each pillar are not determined in advance, and so considering information problems and normative values and goals. Showing these three goals separately at first emphasises their independence and at the same time addresses the interdependency - complementarities and conflicting goals.

Accepting this concept the long-term social development not only depends on natural or ecological capital stock but also on social capital stock. Economic development without intergenerational allocation conflicts is ensured only when at least one of these determinants for development increases and/or a substitutability of these stock figures may be presumed, or when innovations are able to surpass bottlenecks which result when one of these factors will impede the development.

Innovations are assigned a central role in aspects of sustainability. They are assumed to be key factors for the solution of many conflicts and for the mobilisation of synergies between environment, economy and society. The definition of innovations used here is a very comprehensive one[3]. The notion of economic innovation (changes in production process, products and organisation) is enlarged, as besides technical and economical development social (life style and consumption patterns) and institutional innovations (general frameworks of society) are included as well. Concerning innovation phases, not only market introduction but also the first adoption are regarded as well as the phases of invention and diffusion. Sustainable innovations analogous to the three pillar model are not only innovations which lead to a better achievement of environmental objectives but at the same time to innovations with positive effects on economic and/or social goals.

Innovations are regarded as the result of activities in innovation systems. According to Metcalfe (1995), a national system of innovation is regarded as a set of distinct institutions which jointly and individually contributes to the development and diffusion of new technologies and which provides the framework within governments' form and implement policies to influence the innovation process. As such it is a system of interconnected institutions to create, store and transfer knowledge, skills and artefacts which define new technologies.

2.2 Sustainability and Product Related Waste Management

As a result of the UN-conference in Rio de Janeiro in 1992 the importance of the vision of sustainable development increased for the waste policy too. In the Agenda 21, in connection with product related waste management policy for sustainable development, especially chapter 4 "changing of consumption patterns" as well as chapter 21 "environmentally sound management of solid wastes"[4] are relevant. Concerning changing of con-

3 This definition is based on the results of the BMBF-Forschungsverbund „Abschätzung innovativer Wirkungen umweltpolitischer Instrumente (FIU)“ and is as well the basis of the :[riw]-framework (Klemmer et al. 1999).

4 See United Nations Division for sustainable Development (2001) Agenda 21, chapters 4 and 21.

sumption patterns the goals "optimisation of the use of resources by increase of energy and material efficiency" are named. With respect to the last point environmentally friendly products should be introduced as well and in the production and consumption sector old products should be used and recycled. For the "environmentally sound management of solid wastes" measures of waste prevention should be given highest priority especially through the reduction of non-sustainable production and consumption patterns, additionally re-use and recovery of waste should be forced as well. Environmentally sound treatment and disposal of waste have the lowest ranking. With the vision of sustainable development the causal relationship of waste accumulation and the total material input in an economy including primary resources is emphasised.

Sustainability-vision concretised in the Agenda 21 had consequences on the European and German waste policy. This becomes evident in the case of the extended objectives of the German closed substance flow and management act, the codification of the principle of product responsibility and in the case of the paradigm of orientation at natural circular flows[5]. For instance, the Enquete Commission lines out that measures which are exclusively based on waste legislation and waste policy, are insufficient for a successful sustainability policy (Enquete-Commission 1994). Rather measures are decisive aiming at the prevention of waste. Therefore it is necessary to include ecological design and life and consumption patterns into the concept of a modern waste management. However it already regards this prioritisation to be target-oriented, towards a sustainable development. The EU goes along with the requests of the Agenda 21 requiring a separate consideration of economic growth, utilisation of resources and waste and furthermore requiring policies based on an integrated life cycle concept[6].

---

5 Besides ensuring environmentally sound waste disposal the goal protection of natural resources was included as well. So indirectly the resource aspect of prevention and recovery is considered (§1 Kreislaufwirtschafts- und Abfallgesetz (KrW-/AbfG)), Gesetz zur Förderung der Kreislaufwirtschaft und Sicherung der umweltverträglichen Beseitigung von Abfällen (Kreislaufwirtschafts- und Abfallgesetz - KrW/AbfG) vom 27. September 1994 (BGBl. I S. 2705).

6 Targets and measures of the commission draft of a strategy for sustainable development for the area „use and management of natural ressources and waste " see also 6th Environmental Action Programme as well as the commitments at the summit in Johannesburg.

3 Product related regulations for the Electronics Industry Innovation System

3.1 Directive on waste electrical and electronic equipment (WEEE)

The objectives of the WEEE directive are to be seen in the prevention of waste electrical and electronic equipment, and in the re-use, recycling and other forms of recovery of such wastes so as to reduce the disposal of waste. It also seeks to improve the environmental performance of all operations involved in the life cycle. The directive divides electrical and electronic equipment (EEE) into 10 categories.

The objectives are to be reached by means of a wide range of measures such as

- producer responsibility: producers should take the responsibility for certain phases of the waste management of their products,
- Collection of WEEE from private households: separate collection has to be ensured through appropriate systems, so that private and professional users can return their electrical and electronic free of charge,
- Treatment of WEEE: Member States shall ensure that producers set up systems to provide for the treatment of WEEE.

The directive differentiates between the financing of historic old and new devices which have been put on the market after entering into force of the directive. Only with new devices each producer is individually responsible for the financing with regard to the waste caused by his products. To ensure this he has to provide a guarantee declaration.

Besides that several goals have to be achieved: By December the 31st 2006 a rate of separate collection of at least four kilograms of WEEE on average per inhabitant per year has to be achieved. For recovery and recycling special targets have to be achieved as well by then. Both targets have to be adjusted according to technological development. Member states may set up minimum quality standards for the treatment of collected WEEE; and treatment plants must obtain a permit from the competent authorities. Additionally a number of information obligations will be set. Every producer has clearly to be identified by a mark on the appliance. Producers have to provide re-use and treatment information for each type of new EEE put on the market. Information: to achieve better collection rates and to facilitate recovery of WEEE, users of EEE should be informed about their role in the system. Finally an obligation for monitoring is established. Member states are to provide to the commission information on an annual basis on the quantities and categories of EEE put on the market, collected and re-used, recycled and recovered not at least as a basis to formulate a new collection target.

In Germany the BMU has already published keypoints of future legal provisions on WEEE. In addition to correct transposition of the EC-directives solutions compatible with competition are to be pursued and private responsibility activated. Besides that proven elements of waste management that is proven local authority collection structures should be taken into account. Registration will be in the industry's own responsibility and undertaken by a private law clearing house financed by the industry[7].

With the ranking stated in the objectives the WEEE follows the example of the circular flow economy outlined above. Although this system is a step towards a reduced input of material, however it does only indirectly induce a reduction of material and energy. Above all the goal is an encouragement of concepts and the production of electric and electronic devices grossly including and facilitating their repair, their potential technical upgrading, their re-utilisation, their dismantling and their recycling. As far as material flow is concerned a design for recyclability within the measures for a design for environment is requested.

3.2 Other product related regulations

Besides the WEEE-directive other regulations have or will have an influence on EEE. The directive on the restriction of the use of hazardous substances in electrical and electronic equipment (RoHS)[8] aims to approximate the laws of the member states and to contribute to the protection of human health and the environmentally sound recovery and disposal of waste electrical and electronic equipment. The directive comprises the same product groups as the WEEE-directive except medical devices and monitoring and control instruments. Member states shall ensure that from January the 1st 2006, new electric and electronic equipment does not contain lead, mercury, cadmium hexavalent chromium, polybrominated biphenyls (PPB) or polybrominated diphenyl ethers (PBDE). Exemptions

[7] The future ordinance shall commit manufacturers to organise collection in a central contact office for the local authorities in a coordination office. Both of them should not have any link to the operative tasks. On June 2nd 2003 a "Projectsociety Electro-Old Equipment-Register was founded by the industry associations to set up an according registration and coordination office. Elektro-Altgeräte Register EAR 2004 The EAR-Project http://www.ear-projekt.de.

[8] The European Parliament and Council (2003), EG ABl. L 37/19

from eliminations or substitution or the fixing of new maximum concentration values shall be possible to adapt to scientific and technical progress.

On August the 1st 2003 a proposal for a directive of the European Parliament and of the Council on establishing a framework for the setting of eco-design requirements for energy using products and amending was presented (EuP). The present proposal is the outcome of merging two former initiatives: one on the impact on the environment of electrical and electronic equipment and the other on energy efficieny requirements for end-use equipment. The directive shall establish a framework for the integration of environmental aspects in product design and development to ensure the free movement of energy-using products within the internal market. The directive is in accordance with the principles for the implementation of the new approach as set out in the Council Resolution of May the 7th 1985 to technical harmonisation and standards. Products complying with the eco-design requirements laid down in implementing measures according to this directive should bear the CE marking in order to enable them to be put on the internal market and move freely. The harmonisation is restricted on basic environmental demands on products which are obligatory for all manufacturers. The adherence to these requirements shall be proved by a conformity assessment. And the outward sign shall be the CE marking.

Electrical and electronic equipment will be affected by Integrated Product Policy (IPP) as well. It represents a new approach for product related environmental policy and advocates life-cycle thinking which means that consideration is given to the whole of a product's life cycle from cradle to grave. Furthermore, it seeks to minimise environmental degradation by looking at all phases of products´ life cycle and taking action where it is most effective (design, manufacturing, use, disposal). IPP is flexible as for the type of policy measures to be used, working with the market where possible. The commission wants existing instruments to become more market oriented like environmental management systems, labelling and information concerning the product's cycle. Within the IPP the co-ordination between the measures shall be improved to use synergies. The IPP communication of June the 18th 2003 is part of the Commissions efforts to achieve the goals set down in the EU´s 6th Environmental Action Programme and to fulfil the commitments made by the EU at last year's world summit on sustainable development in Johannesburg[9].

9 http://europa.euint/comm/environment/ipp

3.3 The electronics industry innovation system

The electronics industry innovation system presents itself as a complex system of actors and institutions. Those actors are at first the manufacturers of electronic devices (such as Sony, Philips, Sharp, Miele etc.), their suppliers in the supply chain, e.g. components manufacturers (like Infineon, ECM, AMD, Bosch, Intel), 2nd tier suppliers like the chemical industry or subassembly manufacturers, research and development institutes, technology transfer companies, consultants, banks, recycling and re-use companies, maintenance and repairing service providers, logistics companies, manufacturing devices providers and others. Even the customers in the electronics industry belong to the innovation system since they are directly involved into the innovation process. Above all mentioned actors there are interconnections like "normal" transactions in the relationship between manufacturers and customers as well as institutional arrangements to co-ordinate the innovation process (professional organisations like the EECA (European Electronic Components Manufacturers), BITKOM, ZVEI or R&D networks (ECOLIFE-thematic network). The innovation process is triggered by all these actors. The examples given later demonstrate the systemic character of the innovation process in the electronics industry (see table 1).

4 A Sustainable Innovation Scorecard (SISC) approach for the electronics industry innovation system

4.1 Sustainable Innovation Scorecard (SISC) concept

On the background of the remarks on sustainability and the described legislative context, this chapter sets out the system of indicators called "Sustainable Innovation Scorecard (SISC)" to

- identify the impact of the legislative, the market and societal framework on innovation activities in the electronics industry innovation system,
- evaluate the genesis and implementation of innovations in the electronics industry and their contribution to sustainable development,
- provide a tool for policy makers and innovation actors to enable the development of particular policy measures in this context.

The main objective of the scorecard is to develop a basis for a comprehensive framework of indicators on the impact chain of external drivers -

behavioural change - innovations - sustainability impacts and to serve as a roadmap for finding hypothesis[10].

This framework has the advantage that it does not lead to a closed system of indicators but allows for enlargements by newly developed indicators or indicators which have been less important up to now. For specific questions however, the individual goals, timeframe, and areas will have to be defined. (Clausen und Löbbe 2001; Hafkesbrink 2003). With respect to this background the SISC approach is divided into 3 modules (see figure 1):

- module 1 is based on 'determinants indicators' to evaluate the impact of external innovation drivers on the development and diffusion of innovations ('innovation drivers' or 'driving forces'; question 1 of figure 1). In this section of the indicator system the different drivers are also evaluated according to their selective, dynamic and cumulative impacts on innovation behaviour (refer to question 2 of figure 1),
- module 2 comprises 'evaluation indicators' to describe the innovation and its contribution to sustainable development (question 3 of figure 1). In this section of the indicator system the status of sustainability indicators in the electronics industry innovation system is assessed as well (refer to question 4 of figure 1),
- module 3 contains 'descriptive indicators' to describe a specific innovation or (new) technology according to its position on the life-cycle ('descriptive indicators'; question 5 and 6 of figure 1). In this section of the indicator system 'portfolio matrices' are developed to visualise the sustainability potential of each innovation or (new) technology according to the stage of development, the expected leverage effect on sustainability and expected diffusion barriers. These arguments are compiled to evaluate the "need for action" to further develop each technology to contribute to sustainability (derived from question 5 and 6 of figure 1).

---

[10] The operationalisation of the sustainability aspects of this concept is based on the Driving Force-Pressure-State-Response concept of the UN Commission for Sustainable Development (CSD). In the ongoing debate of operationalisation of sustainability many indicator systems were developed; see UN Commission on Sustainable Development (CSD 2001) as well as EUROSTAT/European Commission (2001) and European Environment Agency (EEA 2001). Especially for the WEEE-problem see European Environment Agency (2003). For existing indicator systems compare the survey in Zieschank (2002).

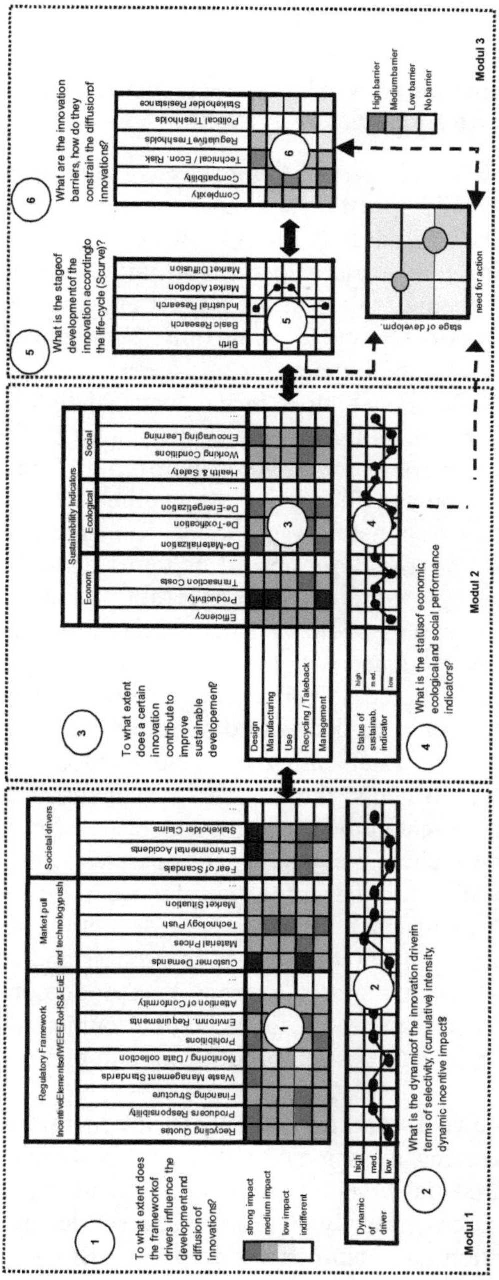

Fig. 1. SISC approach in the Electronics Industry Innovation System (INVERSI)

4.2 Module 1: determinants and their influence on the innovation system

4.2.1 Drivers for innovations in the electronics industry and their influence on the innovation system

The key question of working with this part of the matrix is (refer to question 1, figure 1) to what extent do specific regulatory framework conditions influence the behavioural attitudes of innovation actors in certain fields of action. Linking driving forces and responses of innovation actors to each other, the evaluation procedure to assess the impact of these drivers on the development of innovation is simple using only 4 categories of scores for 'strong', 'medium' and 'low' impact.

Taking the example of the WEEE, which is in the focus of INVERSI and ECOLIFE 2, the take-back obligation obviously influences different areas and functional elements of a company. This results from different legal obligations to be fulfilled, such as collection and recycling quotas, the implementation of the 'producer responsibility principle' and financial responsibility for take-back systems, the definition of certain standards for the waste management and several requirements concerning labelling of products and monitoring of data and mass flows. Manufacturers of electrical and electronic equipment are burdened with the costs of collecting their end-of-life equipment leading to considerable pressure on restructuring the product design (for easy disassembly to decrease disassembly costs), the end-of-life (EOL-) management by establishing new logistical concepts, take-back systems and recycling systems, the innovation management by introducing new environmental oriented requirements like design for environment (DfE) within the supply chain etc. The RoHS operates with prohibitions and restricts the use of certain hazardous substances in electrical and electronic equipment, as for instance lead, mercury and other heavy metals with a considerable impact on the manufacturing process and recycling requirements. The EUE-Directive places a strong burden on companies that produce energy-intensive products to meet environmental requirements and targets in the product's design, production, and end-of-life phase.

Besides the legislative framework conditions there are of course market and technology drivers, influencing a company's strategy. Customer demands, i.e. customer satisfaction, fashions and user requirements rank on top of these drivers placing burden primarily on product and service design but also on corporate information and communication policy. Beyond the product and service price, in the last decade also "green" competitive elements came into place as the result of an increasing awareness of customers for ecological sound products. From the marketing perspective eco-efficiency as a tool to compete on the market was regarded as more and more important, if competition on prices proved to be without effect. “Demand pull” as an indicator to assess the external innovation driver “market” has therefore to reflect the absorbability and demand-elasticity of innovations within the consumer markets, it has to depict customers' behaviour and attitudes regarding problems of sustainability and the willingness of customers to pay for green products. Another driving force are market prices, especially for secondary materials like plastic resins versus market prices of virgin plastic resins determining the absorbability of recycling material from WEEE.

The availability of new technology also plays an important role as innovation driver following the idea that new technologies are driving the products that are created (‘technology push’). For the electronics industry innovation System these technology drivers play a dominant role in the innovation process, such as further development in miniaturisation, upcoming micro-systems and nano-technologies, new wireless communication, technological integration of functions ans.

Another important driver for innovation is the market situation and especially the ‘supply chain pressure’, caused by increased burden placed on the focal manufacturers. Concerning this driver, empirical evidence is given by the fact that tools for supply chain management such as ‘eco-design with suppliers’ are gaining more and more importance since manufacturers have recently been back-shifting compliance duties to their suppliers to a greater extent.

There are also a series of societal drivers with the result of increased awareness of environmental problems among producers and consumers, caused by the public debates following environmental accidents (e.g. Three Mile Island and Chernobyl, Exxon Valdez, Brent Spar), a fear of scandals (like the problem of trans-boarder waste shipments to Eastern Asia) and ‘stakeholder claims’ (such as local, national or international NGO’s activities).

Discussing the link to innovation processes, the following example may be given: on the product design side the challenges to manage, implement and organise eco-design principles are targeted by different external drivers, at least jointly by customer requirements on function and fashion as well as by the actual legislative context. The cumulative impact on design might cause increasing requirements on the design for dismantling (WEEE-impact), to use fewer components or certain materials (WEEE- and RoHS-impact), to use recycled materials (WEEE-impact, green marketing), removal of hazardous substances (RoHS-impact), increased energy-efficiency (EuP-impact), etc.

4.2.2 *Evaluation procedure for determinants as innovation drivers*

On the bottom of module 1 in figure 1 an evaluation of the external drivers is shown. For policy makers on both corporate and governmental level it is important to have a clear view of the dynamics of external drivers influencing the incentive system for innovation behaviour. On this background the SISC approach contains indicators to evaluate these dynamic impacts[11]. Originally developed for the external legislative drivers they can be used for the others as well. These are:

1. *Selectivity*: which is defined as the scope of mandatory obligations covered by a specific regulation influencing certain behavioural corporate parameters. The more selective a mandatory obligation, the more it is directed towards a specific element of the innovation process and the more constraining is its impact on finding corporate solutions to comply with the mandatory obligation. The RoHS is an example for an utmost selective regulation, leading to substitutional technologies or substances. Collecting quotas within a national take-back directive are less selective because it opens a range of solutions from individual manufacturer solution via branch solutions to municipal solutions (Lucassen 2002), thus the innovation process is not constrained as much to a certain direction.
2. *Cumulative intensity*: to understand the impact of a specific environmental policy instrument it is important to learn something about its mutual interrelation with other instruments of the legislative context, even its interrelation with market drivers and technology drivers. On this background cumulative intensity describes the way a regulation works alone or jointly together with other instruments, and if there are reinforcement, acceleration, diminishing or contra-productive effects that increase burden on innovation actors. For instance, with respect to the electronics industry innovation system innovation effects are increased substantially by combining several policy instruments directed to the early design phase of product innovations, placing a heavy burden on alteration of product functions to decrease energy consumption during the use-phase and disassembly- and recycling costs during the end-of-life phase.
3. *Dynamic incentive impacts:* another important indicator to assess the effectiveness of a regulation consists in the extent of its dynamic incentives. It is obvious that a fixed threshold in an environmental directive

---

[11] Indicators "selectivity of legislative obligation" and "dynamic impacts" are based on evaluation criteria of static and dynamic efficiency for environmental policy instruments (Linscheid 1998; Clausen 2000).

rises action to comply with this obligation only up to the threshold value. This can be called a static incentive since there are no dynamic incentives to move beyond the borderline. The indicator "dynamic incentive impacts" is defined as the amount of constant incentive impulses placed on an innovation actor or innovation process to improve (certain elements of) the innovation towards the desired (political) objective. From a sustainability point of view a legislative context should be preferred that produces dynamic incentives as much as possible (supposed the objective is well chosen)[12].

4.3 Module 2: innovation indicators and indicators for their impact on sustainable development

4.3.1 Innovation indicators

A brief analysis of the main characteristics of the present situation within the electronics industry innovation system shows that 3 categories of "innovations" are to be differentiated:

(1) Dialogue and other institutional innovations: In a knowledge-based economy, innovation is a result of a constant flow of both formal and tacit knowledge within an innovation system, it strongly depends on the effective organisation of learning processes by facilitating trustful contacts and interrelations between the innovation players. Especially the genesis and implementation of sustainable innovation requires an extensive co-operation and calls for an intensive dialogue and an open communication platform between the players involved. With these instruments and related institutional innovations the Electronics Industry Innovation System will overcome the problems of co-operation barriers by facilitating interactive learning among different innovation actors.

(2) Strategies and organisational innovations: Since there is always a broad spectrum of solutions to comply with extrinsic regulative drivers or to put intrinsic drivers into practice, the players of the innovation system (re-)act by making plans, by setting up strategies to gain primarily economic and competitive advantages. Against this background the second major 'category of innovations' are instruments of strategy, methodologies and plans to keep in lane with necessary requirements. In this respect, strategies are more than the application of a software tool, it is also the or-

---

12 The :[riw]-program of BMBF contains interesting case studies both on timing aspects as well as on dynamic incentives impact of environmental policy instruments, for further details see http://www.riw-netzwerk.de/

ganisational and institutional setting in which the application of tools takes place.

(3) Tools, products and technologies: The third 'category of innovations' are the most visible ones, that means the instruments covering a certain scope of problems of sustainability like evaluating the eco-efficiency of a certain process. It has normally a limited scope as it is clear that a tool to support decisions on economic, technological, ecological or social measures cannot cover the complexity of the interdisciplinary context. So it is obvious that the Electronics Industry Innovation System needs a toolbox with manifold instruments in place to help innovation actors to move towards sustainability. This toolbox may also be the most critical one as the outcome of decision tools is very important to produce reliable and valid information on the road to sustainability.

Since innovation in the Electronics Industry takes place in the whole life cycle of electronic products, the three main categories *dialogue*, *strategy* and *tools* have to be explicitly embedded into the main stages of the life cycle of an electronic product, i.e. *design*, *manufacturing*, *use*, *end-of-life* and *management*, understanding "management" as a cross lateral task accompanying the life cycle.

Figure 2 gives an overview about the conceptual framework of the descriptive indicators:

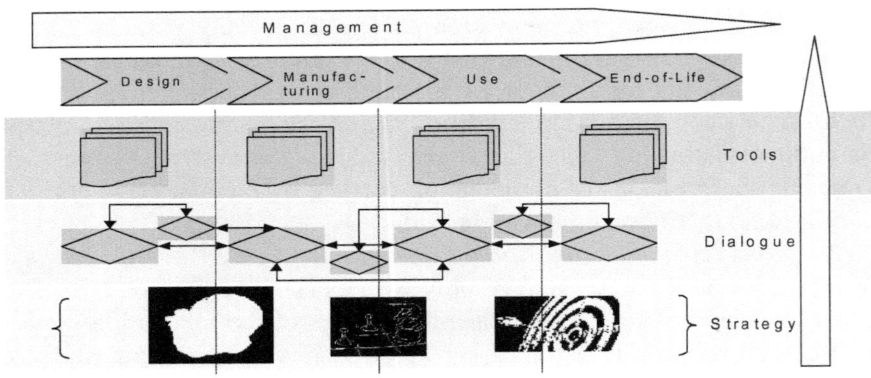

Fig. 2. Conceptual framework for descriptive innovation indicators in the electronics industry

Table 1 depicts a selection of 45 out of 120 single innovations, rated as most important for the Electronics Industry within the ECOLIFE 2 network:

Table 1. Innovations in the electronics industry innovation system regarded as most important[13]

	Design	**high**	**med**	**low**		**Use**	**high**	**med**	**low**
Dialogue	Ecological idea dissemination through the supply chain	9	5	4	**Dialogue**	Customer information and education on usage	11	5	3
	Eco-design with suppliers	9	6	4		Communication of products impacts to the consumer	9	5	4
	Management of eco-cost reduction with suppliers in manufacturing & design	7	7	4		Understanding customer behaviour and communication with customers	10	5	2
	Communication strategies among companies	7	6	4	**Strategy**	Energy efficiency in use	12	3	4
	Information dissemination to SME	7	6	4		New business models (leasing etc.)	9	5	4
						Recovery / EOL	**high**	**med**	**low**
Strategy	Design for environment	14	4	2	**Dialogue**	Information communication between electronics industry and recyclers	12	2	3
	Design for chemical content	4	10	5	**Strategy**	(Cost effective) EOL and recycling technologies	13	5	1
	Design for EOL, dis/assembly	10	5	3		Standards and technical specifications for recycling	8	8	2
	Integration of DFE in conventional management systems	8	6	5		Logistical concepts concerning collection of used electronics	10	8	1
	Substitution of hazardous materials (e.g. BFR, VOC's, semi-conductors) in products	10	9	0		Market development for recyclates	9	8	1
	Renewable materials	5	10	4	**Tools**	Disassembly analysis	5	10	4
						Automatic disassembly technologies	7	7	5
Tools	LCA/LCC including simplified LCA	7	8	5		Recycling of materials and components, special interest: PCBs, copper/glass, packaging materials	7	8	5
	Database on materials/components for DFE	11	5	4		Development of (public) take-back schemes for EOL	12	3	3
	Life cycle engineering	8	6	5		**Management**	**high**	**med**	**low**
	New substrates for PWB	3	13	4	**Dialogue**	Supply chain management	11	5	1
	Halogen-free flame retardants	6	10	4		Knowledge management, knowledge transfer and distribution	9	6	2
	New flame retardants materials	10	5	3		Education and training	9	7	1

13 This table comprises the result of a technology experts dephi (32 experts), conducted in 2003 in the ECOLIFE thematic network (see for details, ECOLIFE II - Eco-efficient life cycle technologies - state-of-the-art technology report in the electronics industry innovation system (Hafkesbrink et al. 2003)).

Table 1. (cont.)

	Manufacturing	**high**	**med**	**low**	**Strategy**	Legislation monitoring of RoHS, WEEE, IPP, EEE etc.	8	7	3
Dia-logue	Dissemination of best industrial process	10	7	1		Ensuring legal compliance	7	6	3
Strategy	Substitution of hazardous materials in production	10	7	2		Green strategy making and green innovation management	6	7	5
	IPPC	4	9	3	**Tools**	Consultation between industry and government	9	5	5
	Improved manufacturing of materials, components & subassemblies	4	10	4		Roadmaps, performance measurement			
Tools	Lead-free soldering	12	4	3		Eco-mapping of drivers (government, customers, NGO's)	4	9	4
	Eco-efficiency of manufacturing	7	7	5		Ensuring legal compliance of suppliers	6	9	3

4.3.2 Indicators for the evaluation of impacts on sustainable development

Operationalisation of the three pillar concept

Accepting the three pillar concept as operationalisation of sustainability indicators have to be defined and verified which not only describe the stock of ecological but at the same time the stock of economical and social capital or as this is hardly ever possible measure the losses of substance or investments into these stocks. Such abstract principles are not directly applicable for the analysis of concrete questions. So a specific innovation system like the one of the Electronics Industry needs a certain break-down of these principles into specific and operational indicators. These indicators are manifold and have at least – according to the three pillar concept to provide information on the impact of innovations on economic, ecological and social improvement.

Ecological improvement: These indicators are defined in relation to the use of natural resources and capability of the environment to absorb emissions. The connections between the economic system and the environment can be expressed by:

- de-materialisation, i.e. a certain positive relation of material input and output,
- de-toxification, i.e. a certain decrease in the amount of hazardous substances,

- de-energisation (or de-carbonisation), i.e. improvement in the relation of energy input and output (energy-efficiency).

Economic improvement: The indicators chosen here do not only show the consequences for economic variables but express a linkage to ecological variables and make an evaluation between both objectives possible at the same time. This includes:

- efficiency, i.e. a certain positive relation of costs and benefits in monetary terms,
- productivity, i.e. a certain positive relation of output and input in quantitative terms.

Social improvement:

- encouragement of learning and education in the society,
- improvement of job security of employees involved,
- improvement of health security of employees involved and of applicants.

These indicators have to be broken down to describe the effects of an electronics industry innovation system. Empirical evidence on the ecological and economic effects of especially the WEEE within the Electronics Industry Innovation System is given in the following chapters.

Ecological aspects

Within INVERSI several investigations have been carried out (evaluation of data collections and literature, visiting of conferences, leading of interviews) to gain empirical evidence on indicators of ecological improvement. That means an operationalisation with respect to the goals of the WEEE-directive. Up to now only a few quantitative indicators are available allowing an assessment on an aggregate level (European Environment Agency 2003). In the future, the monitoring system of the directive will give the necessary information.

The empirical results show that in future a considerable improvement in ecological sustainability is supposed to be achieved. Based on the trends of the past the following development might be expected for the future when the WEEE-directive is implemented:

(1) Reduction of waste amount: in the EU in 1998 about 6 million tons of electronic waste have been disposed, most of them via landfilling. The total amount of WEEE generated in the EU is estimated at 6.5 to 7.5 million tons per year in the late 1990s increasing by 16 to 28 % every five years. The collection of WEEE in selected European countries is depicted in figure 3.

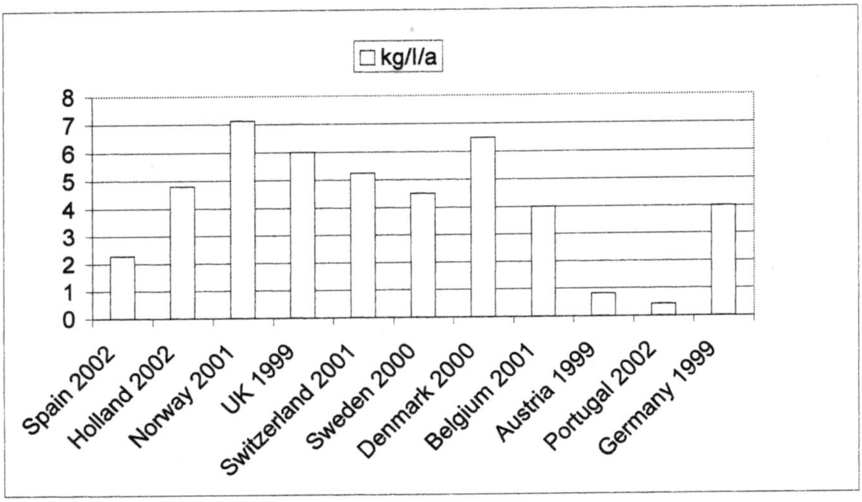

Fig. 3. Collection of WEEE in Europe in kg per inhabitants per year (source: own compilation)

With a collection goal of 4 kg/I/a about 1.5 million electronic devices are separated from the existing waste disposal. Various estimates of the quantity of WEEE indicate that the collection target of 4 kg per inhabitant constitutes only 25 % of the overall annual generation of this waste. The collection results obtained so far, however, may not supersede the fact either that this volume is usually achieved with the collection of white goods, as TV's, monitors etc. Most of the small electronic devices such as mobile phones and in particular devices fitting in normal garbage cans are still disposed with the municipal solid waste. As a result, the recycling quotas for the single categories obtained so far are evaluated as partly problematical. If this does not improve the recycling quotas the directive may not be fulfilled.

(2) Increasing material productivity: Numerous product examples of the big players like Sony, Phillips and Electrolux today are commercially exploited with green arguments. For example, each new product generation of mobile audio and video devices (for instance Walkman, Handycam, Discman) is smaller and lighter. In addition, by replacing hardware with software and e-solutions an increasing de-materialisation takes place, substituting physical products with electronic.

(3) Material substitution: In ecological matter, the existing strategies of material substitution contain both positive qualitative and quantitative effects. Material substitution does not only mean the substitution of hazard-

ous with non-hazardous substances (lead-free solders, displacement of toxic developers and fixative solvents, substitution of solvent-containing cleaners in copying machines etc.) but also the replacement of heavy with lighter materials (for instance optimisation of the counterweights in washing machines, development of flat screens, etc.). This strategy could change the problem from a quantitative to a qualitative one.

(4) Re-use strategies for the re-use of electronic devices or components directly contribute to close the loops on a high utilisation level. In fact, there is an evolving market for the re-use of electronic devices and components in Germany. Computers placed out of industrial service are handed on to schools and other social institutions. On the internet market place eBay, approx. 500,000 auction offers for used devices from the areas audio, electronic devices, TV, video and electronics are constantly to be found. The actual size of the re-use market can hardly be measured, reliable data and information are missing. In selected market segments, specialised market participants are operating to exploit re-manufactured mobile phones, PC's, single modules and components, offering their products and services (spare part services etc.) in a world wide context. Companies like Kodak, IBM and Hewlett Packard have been running concepts of re-manufacturing for years following economic arguments, for instance in the area of copying machines (re-use of parts, such as ventilators) and servers (IBM). The re-filling of toner cartridges belongs to one of the standard operations running a copying device today.

(5) Quality of recycling: based on experts' view – the required quotas of the WEEE-directive will be achieved without any problem in the medium to long run with technical improvements of recycling. The recycling industry has considerably improved the recognition and separation technology shifting to semi-automated processes and thereby improved economic efficiency leading to a double dividend in environmental and economic matters. But, according to the state-of-the-art of recycling technology, not all of the extensions in sorting and disassembly depth are supposed to be eco-efficient, i.e. the additional benefit resulting from a more in-depth disassembly must be paid with a substantial cost enhancement.

(6) Avoidance of hazardous substances: The still largest problem in the recycling is the recognition and separation of plastics, due to approx. 60 different kinds of plastics, the incorporated flame retardants, other additives (pigments, stabilisers etc.) and other contaminations (labels, foam, metal foils etc.). In addition, low prices for primary materials are restraining the development of a stable secondary raw material market and the exploitability of secondary raw materials is limited as the result of further restrictions (e.g. standards and regulations).

However, especially after the RoHS implementation, the amount of hazardous substances polluting the environment will decrease. This concerns in particular cadmium, lead and mercury e.g. from batteries, from fluorescent tubes, toxic organic compounds from liquid crystal displays, polychlorated biphenyles e.g. from condensers, fluorine chlorinated hydrocarbons from coolers, flame retardants and heavy-metalliferous additives from plastics, gallium arsenide from light emitting diodes up to asbestos in older household appliances.

(7) Life-cycle oriented manufacturing and product strategies: Most manufacturers have recognised that with product innovations a life cycle perspective is important, to consider for instance improvements in end-of-life phases regarding their effects on other phases of the life cycle. This becomes more and more important, if one considers that the environmental effects of electrical appliances result on average only to about 2-5% from the end-of-life phase, to 10-35% from production, to 5-15% from packing and transport, but to 50-80% from the use phase. At the same time, by making more and more use of LCA Tools (life cycle analyses), more profound decisions on the design of new products or product changes will be obtained, e.g. with respect to energy.

(8) New utilisation strategies: The implementation of new utilisation concepts and strategies (multiple use, community use, use cascades, leasing, use instead of possession) requires a by far more comprehensive innovation development. Here new thinking of all market actors is required. Manufacturers have to re-think business processes to shift earning possibilities from "old economies strategies" (earnings as a result of shortening the innovations cycle) to "new sustainable economies strategies" (earnings as a result of life time extension, energy minimisation, intelligent services, etc.). The users must reorient their opinion that it is less important to be the owner of a product (possession-thinking), than to buy and use its functions (need satisfaction). The diffusion of such system innovations in consumer segments, however, is only at the very beginning.

Economic aspects

Estimations of the ZVEI number the compliance costs of manufacturers and consumers in Germany, resulting from the WEEE and the RoHS, on altogether approx. 350 to 500 million €., whereas the costs per equipment range from 8 € for a washing machine, to 10 € for a television set up to 15 € for a refrigerator (O.N. 2002).

As a central condition for efficient take-back systems economies of scale play a dominant role. Achieving them is crucially important for all investments in collecting and transportation logistics, recycling plants and also for the market development for the exploitation of secondary raw materials. Missing economies of scale in reverse will be the central innovation obstacle for the implementation of the WEEE, in particular by rising costs per inhabitants and year mainly in geographical regions with small population density.

In the discussion about the WEEE implementation it became evident that the economic efficiency of take-back logistics obviously depends strongly on the form of the implemented take-back logistics structures. Thus Hewlett Packard (Hieronymi 2002) reports that the Dutch and Swedish system of take-back causes smallest costs due to the fact that all services of the take-back and recycling are contracted directly between manufacturers and assigned companies. Pool or consortia solutions however cause take-back costs which are in comparison to the above mentioned up to 18 times higher. To that extent industry prefers WEEE implementations with recyclers and take-back companies being in constant competition among each other.

The economic efficiency of the implemented system has also to be measured against the sustainability of the incentive mechanism for ecological product improvements. Thus at least during the first years after the implementation of the WEEE-directive when having to handle historic old devices the problem exists[14] in all countries. Likewise the problem of handling orphean products exists, products of which the producer does not exist anymore or as in the case of crossborder sales cannot be addressed (see INVERSI).

To that extent the economic efficiency of the financing systems will have to be measured against the circumstance whether cost categories exclusive or at least predominantly influencable by the manufacturer categories are to be found in the financing systems. In view of the described transition problems this is to be questioned.

Estimations on the cost volume of abandoned old devices and freerider products are only very indistinct for Germany and are reported to be in the range of three-digit million amounts (Skottheim 2002) On this background the industry deplores the fact that collective collecting systems with cost agreements oriented at market shares etc. scoop out the incentive functions since no more direct links between their own products and the cost volume exist, and if orphan products must be paid for in addition.

---

14 i.e. the producers of which are no longer active on the market or cannot be identified.

4.4 Module 3: Descriptive Indicators for innovation-stage of development and derivation of "Need for action"

As shown in Module 3 of fig. 1 an assessment of the stage of development of a certain technology and of possible innovation barriers is asked for. This part of the indicator system is directed towards a particular evaluation of research- and technology policy as well as environmental policy related questions. The basic assumption in this context is: When a technology/innovation is evaluated according to its position on its life cycle, the leverage effect on improving sustainable development and diffusion barriers can be examined. After that the "need for action" concerning policy and corporate measures can be derived to improve its contribution to sustainability. On this background the concept of "Need for Action" comprises 3 evaluation steps:

1. Defining the stage of development of an innovation according to the life-cycle,
2. Defining the leverage effect of an innovation to contribute to sustainability,
3. Analysing the diffusion barriers of innovation.

The compilation of these evaluation steps is edited within portfolio matrices to provide an overview about the "need for action" within one integrated approach.

4.4.1 The evaluation of descriptive indicators

Ad (1): stage of development: to define the stage of development of a specific innovation or technology, the SISC concept is based on a technometric approach as a method to determine primarily the technological and economic performance of technologies, and to define, ascertain and process performance indicators of technologies (Hafkesbrink and Krause 1995). Indicators to evaluate the position of a technology according to the stage of development are in practice (Hafkesbrink et al. 1993):

	State of the art	Pace-setter technologies	Key-technologies	Basic-/displaced technologies	
Mapping of EEE technologies	Performance				
	Stage of development	Birth	Growth	Maturity	Age
Indicators for evaluation	Technical risk	High	Medium	Low	None
	Budget required	Medium (basics)	High (application)	Low (cost reduction)	Very low
	Entry barriers	R&D Potential	Availability of know how, services, components	Investment costs for facilities, licenses etc.	Application know how
	Development requirements	Scientific basis	Industrial basic research	Applied research	Cost oriented development

Fig. 4. Examples of indicators to assess the stage of development

The life-cycle view is usually related to the following stages:

- Birth (the technology is only to be found in scientific basic research)
- Growth (the technology is yet in industrial or applied research)
- Maturity (the technology is almost adopted and/or diffused in industry)
- Age (the performance of the technology has almost reached its peak).

Ad (2) Assessment of Leverage Effects: The leverage effect of a particular technology or innovation on the sustainability indicators may be assessed by analysing the 'sum' of the scores given in Module 2 of figure 1 with respect to the different sustainability indicators. This follows again a simple evaluation procedure, using a non-metric qualitative scale of just "adding" the individual scores. More sophisticated evaluation procedures are using a complex quantitative or even a metric scale by assigning quantitative values of benefit to the scores "high-medium-low". Other evaluation procedures are referring to the relative or absolute alteration of sustainability indicators to assign the leverage effect as 'high', 'medium' or 'low'. However, a normative vision or value judgement has to be established for each individual innovation linking to each individual sustainability indicator and the evaluation procedure has to be kept constant over time and constant across different sustainability indicators to allow for inter-subjective comparability and comprehensibility.

Since evolving innovations or technologies, which might be positioned in the stage of basic or applied research and whose contribution to sustainability is not yet provable on an ex post basis are also referred to, the "potential" of a technology/innovation to contribute to sustainability has to be defined. It is referred to as the principal capability of a technology to contribute to sustainable development based on most likely prospective expectations.

Ad (3) Assessment of Diffusion Barriers: The structure of the diffusion criteria used in the SISC approach relies on earlier works by Rogers (2003) and Schoemaker (Rogers and Shoemaker 1971) defining the "diffusion" of an innovation as the process by which an innovation is communicated through certain channels over time among members of a social system. According to these works, the key indicators to explain the diffusion of an innovation are relative advantage (degree to which an innovation is perceived as being better than the idea it supersedes), compatibility (degree to which an innovation is perceived as being consistent with existing values, past experiences and needs of the potential adopters), complexity (degree to which an innovation is perceived as difficult to understand and use), trialability (degree to which an innovation may be experimented with on a limited basis) and oberservability (degree to which the results of an innovation are visible to others)[15]. Following especially the evolutionary analysis of technical change (Nelson and Winter 1982), the set of diffusion indicators has to be extended to cover also the barriers in the institutional framework such as regulative or political thresholds as well as stakeholder claims. As empirical references some examples for these diffusion indicators will be given[16]:

(1) Complexity: some of the technologies / innovations mentioned in table 1 are narrow in scale and scope (like a particular plastic separation technology), others are complex because as a pre-condition for their diffusion, an extensive alteration of business structures, management procedures and infrastructure conditions has to be worked out in the innovation system (for example: new utilisation concepts and product-service shifts require a complex co-operation of various innovation actors along the supply chain involving additional players from service and maintenance).

(2) Compatibility: Take-back concepts to fulfil the requirements of the WEEE that are compatible with the current (municipal) systems of collection and management are more likely to disseminate than those, which are not compatible.

(3) Regulative thresholds: competition conformity, conformity with the GATT treaty, with WTO standards, with ISO norms are regarded as the most important regulative thresholds in the implementation of take-back obligations. The German Antitrust Division (Kartellamt) so far felt impelled to intervene several times within the implementation of the German

[15] http://hsc.usf.edu/~kmbrown/Diffusion_of_Innovations_Overview.htm

[16] see again figure 1

Packaging ordinance due to deficits in the institutionalisation of the Duales System Deutschland as the systems operator.

(4) Political thresholds: in scientific discussions environmental policy in Germany is regarded as a new governance concept based on consensus and huge participation of enlisted actors. Under these circumstances the implementation of take-back ordinances is carried out in a complex institutional framework of professional associations, NGO's, action groups of various kinds, representatives of large scale companies etc. The political thresholds for the implementation of a particular take-back system may be a difficult task to do, since the principle of consensus might lead to long-lasting time windows concealing opportunistic strategies of one or the other party involved.

4.4.2 Portfolio matrices

The final evaluation step in the SISC approach is composed of a compilation of the preceding steps. To use the '"potential" when diffusion barriers are high the need for action has to be derived at. To operationalise the indicator 'need for action', the items depicted in Fig. 5 are used within the evaluation procedure. According to these 'evaluation rules', the innovations/technologies are rated with the help of different criteria assuming that there is

- only a low 'potential' for those existing technologies, which have already reached their peak performance, or for those new technologies with still no economic feasible way to implement it on a corporate level.
- a medium potential for those existing technologies which promises further value added for sustainable EEE.
- high 'potential' for those technologies, which promise a maximum of added value for sustainable development, which have not yet utilised their abilities.

As can be seen from the criteria and indicators used, the evaluation procedure maybe called "qualitative" since non-metric scales for these indicators are applied. As a result, the evaluation of technologies maybe called an "estimation" based on the experiences of a balanced mix of expertise from industrial and research actors of the Electronics Industry Innovation System. However, an extensive learning process was facilitated with the help of these indicators in the process of working out a consensual and mutual understanding of the impact chain 'driver – innovation – contribution to sustainability'.

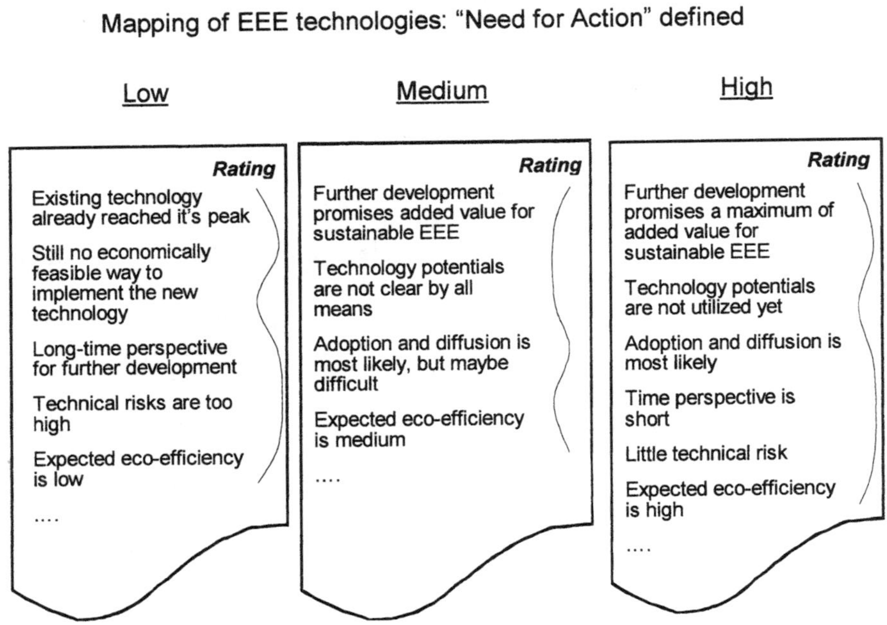

Fig. 5. Evaluation indicators for „Need for Action"

Module 3 of the SISC approach then comprises the evaluation results concerning the state-of-the-art technologies and the need for action within integrated portfolio matrices. These 'Portfolio Matrices' are set up to visualise the sustainability potential of each (new) technology according to the stage of development and the need for action to further develop each technology to contribute to sustainability. This also refers to the technometric background of the SISC approach. To illustrate the position of a technology, portfolio matrices are used with two evaluation axis, where the first axis denotes the stage of the technology according to birth, growth, maturity and age and the second axis denotes the "Need for action" (NfA) to further develop the technology to contribute to sustainable development (SD). It can be interpreted in the following way:

- high NfA: the technology promises to contribute to SD very much, adoption of industry is most likely, eco-efficiency is high etc.
- medium NfA: technology potentials are not clear by all means, expected eco-efficiency is medium etc.
- low NfA: existing technology that already reached its (it's) maximum ability, expected eco-efficiency of new technology is low.

Empirical examples of fig. 6 may be demonstrated with the help of different Technologies of Design within the ECOLIFE communication process[17] leading to considerable insights into the sustainability evaluation of present technologies in the Electronics Industry. In order to retain the distinctness of the approach, in this paper only the final steps of evaluation are presented.

In the area of design/dialogue (Hafkesbrink 2003)[18] a set of key drivers related to the implementation of eco-design within the supply chain has been identified as follows:

- Large-scale manufacturers have introduced eco-design as a tool since the early 1990s as the result of upcoming legislation, indicated by the draft German ordinance on take-back obligation of 1991 (Hafkesbrink et al. 1998) Up to now there is a considerable amount of experiences in that area[19]. However especially SMEs in the supply chain are not aware of the forthcoming legislation and for those who are it is not well understood. Manufacturers are beginning to communicate the requirements of these directives related to eco-design but only to those suppliers high up in the supply chain.
- Because manufacturers are back-shifting the requirements to their suppliers, supply chain pressure is more of a driver than legislation at the current time.

---

[17] Thanks especially to Leigh Holloway (ProActus, UK), Glenn Johansson (ivf, Sweden), Martin Charter (cfsd, UK), Emanuela Scimia (febe, Italy), Constantin Herman (IKP, Germany), Ab Stevels (Phillips, Netherlands), who have fed in their experiences on eco-design into the evaluation.

[18] See again figure 5

[19] See the comprehensive eco-design guide developed in ECOLIFE 1 (http://www.ihrt.tuwien.ac.at/sat/base/Ecolife/ECOIndex.html)

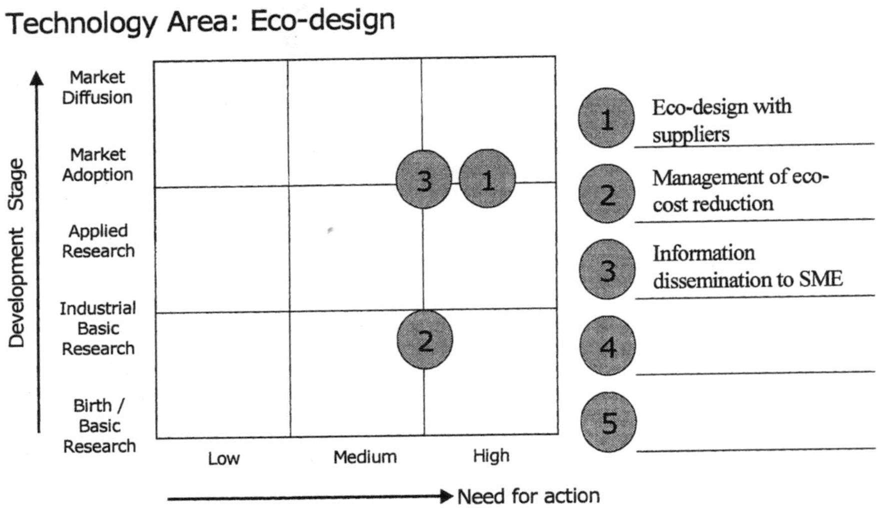

Fig. 6. STS evaluation of dialogue technologies in the area of eco-design

On this background for positioning of "eco-design with suppliers" within the SISC matrix the following issues seem crucial:

According to the stage of development the "technology", i.e. principal measures of 'dialogue to suppliers' and 'tools for benchmarking of suppliers'[20] have already reached industrial application (first adoption within case studies) though a broad diffusion of the technology has not happened up to now. The need for action to further develop and disseminate the technology for sustainable development in the electronics supply chain however proves to be high: What is obviously needed are more industrially relevant case studies on eco-design supply chain management, case studies that not only show the results of eco-design activities but steps that were taken and tools that were used to achieve these results. At this time there is a lack of good industrial examples but this should change within the next 2 years. With the deadlines for legislation coming closer regulation will begin to play a more important part in eco-design but at this stage the supply chain is still the most effective driver.

It will become necessary that large companies engage their suppliers and customers in projects that offer support, information exchange and the chance to network. To develop truly 'eco-designed' products it is essential that all players within the supply chain are involved. The key needs to develop and promote eco-design in the supply chain are:

20 See again figures 2-4.

- Better communication between customers and suppliers,
- The development of eco-design standards and requirements along the supply chain,
- Eco-design requirements to become part of the standard supply chain agreements and contracts,
- Easy to use tools and methods that do not require specialist expertise and that especially integrate the supplier's environmental performance to the purchase turnover,
- Clear industrially relevant case studies on eco-design supply chain management.

Improvement of dialogue and tools in the area of eco-design with suppliers will substantially improve the greening of the entire supply chain, especially if a linkage between environmental performance and purchase turnover will be implemented within tools or benchmarks and move the supply chain significantly towards sustainability. High eco-efficiency effects will be attainable because a reduction in environmental load of parts and components will correlate with price reduction in purchased goods. A feasible value added for sustainable EEE will be realised, at least within the two dimensions of economic and ecological improvement. Because of the double dividend aspects, a broad diffusion seems to be possible within a relatively sizeable timeframe. Social improvements may occur if reduction of material resources within the supply chain also includes hazardous material, which is clear for all material tackled by the RoHS.

The stage of development of the technology "Management of eco-cost reduction" is more or less to be located in theory and in basic research. Applicable tools are not yet developed. The need for action is to be scored as medium to high, since it makes sense to spread the idea of double dividend mechanisms throughout the supply chain by linking the resources consumption in manufacturing processes to the corresponding environmental load with purchase negotiations. Problems may occur in data accuracy and reliability according to the mass balance of suppliers necessary to evaluate the amount of resources decrease possible and the corresponding potential of cost reduction. Thus the chances for implementation of such tools may be ambiguous. In this field, additional research and case studies seem to be necessary.

For "Information dissemination to SMEs" the stage of development should be classified as "market adoption", since a lot of tools (either web-based or "classic information and decision support systems) are available, though they might be far away from a broad diffusion to, and continuous recognition by, SMEs. The need for action may be qualified as high, since a more intensive recognition of environmental related information through

SMEs is crucial. From the experiences with eLCA it seems that further development necessary for dissemination does not definitely mean more information but tailor made information according to the special situation of SMEs.

5 Conclusions

An essential aspect of INVERSI is to first point out dimensions and direction of innovations which will be induced by the WEEE-directive as important changes of institutional conditions in interaction with other relevant determinants and then evaluate these innovations with regard to their impacts on sustainable development. As a comprehensive indicator approach SISC was used to show the contextual relations. It proved to be useful to show the influence of environmental regulations on the different areas of innovations (design, manufacturing, use, end-of-life) and to show the direction and intensity of the regulation impacts. The indicator system was also helpful evaluating the impact of innovations with respect to the achievement of the goals of the WEEE and for isolating the influence of the directive from the influence of other regulations. Thus it could serve as a roadmap to develop hypotheses concerning the efficiency of different framings of such an instrument. These hypotheses could be further evaluated in interviews. The SISC approach stresses the linkage between setting up framework conditions and micro-economic reactions and focuses more the microeconomic aspects. So, concerning INVERSI it has to be complemented by other indicator sets and methods.

The SISC approach proved to be ideal for a project like ECOLIFE 2 as this project can serve as an example on how to instrumentalise tools along the impact and evaluation chain from (macro) legislative framework conditions to (micro) (re-)actions on a corporate level. In this context definitions of indicators and their contents are especially important because the actors involved do not share the same cultural background, are using different systems of evaluation reference according to their interests. Especially ECOLIFE 2 used these indicators as a communication tool for experts interviews, to structure the research process and to concentrate on important issues of the research. It thus initiated fruitful discussion processes between business and science. The chain of analyses and evaluation steps within the SISC approach primarily provides the basic scheme for a scientific hypothesis system for further research. It shows itself as a pragmatic approach, which on one side is a precondition for studies concernig electronics innovation system on the other side it is going on to develop during working on such projects.

6 References

Bundesministerium für Umwelt, Naturschutz und Reaktorsicherheit (BMU) (2003) Eckpunkte künftiger Rechtsvorschriften zu Elektro- und Elektronikaltgeräten in Deutschland. Berlin

Clausen H (2000) Rücknahmeverpflichtungen als Instrument der Abfallwirtschaftspolitik. Münster

Clausen H, Löbbe K (2001) Indikatoren nachhaltiger Entwicklung – einige Anmerkungen aus der Sicht der empirischen Wirtschaftsforschung. In: Eckey HF, Hecht D, Junkernheinrich M, Karl H, Werbeck N, Wink R (eds) Ordnungspolitik als konstruktive Antwort auf wirtschaftspolitische Herausforderungen. Stuttgart

Daly HE (1990) Towards some operational principles of sustainable Development. Ecological Economics 2: 1-6

Calvo A, Rahrig K (1997) Diffusion of innovation http://hsc.usf.edu/~kmbrown/Diffusion_of_Innovations_Overview.htm

Elektro-Altgeräte Register EAR 2003 The EAR-Project http://www.ear-projekt.de

European Parliament and Council (2003) Directive on Waste of Electronic and Electrical Equipment (WEEE) (directive 2002/95/EG). EG ABl. L 37/24

European Parliament and Council (2003) Directive on the Restriction of the use of Hazardous Susbstances in Electrical and Electronic Equipment (RoHS) (directive 2002/95/EG). EG ABl. L 37/19

Ferrendier S, Mathieux F, Rebitzer G, Simon M, Roelich D (2002), ECOLIFE 1: the comprehensive Eco-design Guide http://www.ihrt.tuwien.ac.at/sat/base/Ecolife/ECOIndex.html

Enquete-Commission „Schutz des Menschen und der Umwelt" of the German Parliament (deutscher Bundestag) (ed) (1993) Verantwortung für die Zukunft – Wege zum nachhaltigen Umgang mit Stoff- und Materialströmen. Bonn

Enquete-Commission „Schutz des Menschen und der Umwelt" of the German Parliament (deutscher Bundestag) (ed) (1994) Die Industriegesellschaft gestalten. Perspektiven für einen nachhaltigen Umgang mit Stoff- und Materialströmen. Bonn

European Environment Agency (EEA) (2001) Environmental Signals 2001. European Environmental Agency regular Indicator Report. Environmental Assessment Report 8. Copenhagen

European Environment Agency (EEA) (2003) Waste from Electrical and Electronic Equipment - Quantities, Dangerous Substances and Treatment Methods. Copenhagen

EUROSTAT/European Commission (2001) Measuring Progress Towards a More Sustainable Europe. Proposed Indicators for Sustainable Development (Data 1980-99). Luxembourg

Hafkesbrink J, Krause M, Mokwa W, Rospert M (1993) Technologien für Mikrosysteme – Stand und Entwicklung, Düsseldorf, p. 157; Arthur D. Little (Ed.), Management der FuE-Strategie, Wiesbaden 1991, S. 68.

Hafkesbrink J, Halstrick-Schwenk M, Löbbe K (1998) Abschätzung der innovativen Wirkungen umweltpolitischer Instrumente in den Stoffströmen Elektroaltgeräte/Elektronikschrott. Untersuchungen des Rheinisch-Westfälischen Instituts für Wirtschaftsforschung (RWI) 26. Essen

Hafkesbrink J, Herman V, Kopacek B, Brandstötter M (2003) ECOLIFE II - State-of-the-Art Technologies in the Electronics Industry Innovation System. http://www.ihrt.tuwien.ac.at/sat/base/EcolifeII/ECOLIFE%20%20II%20D7%20summary%20FINAL.pdf

Hafkesbrink J, Krause M (1995) Technometrics as a Tool in the Evaluation of Innovation-Support-Programmes - Using the Example of Microsystems Technologies (MST). In: Becher G, Kuhlmann S (eds) Evaluation of Technology Policy in Germany. Dordrecht

Hieronymi K (2002) Norway, Switzerland, Netherlands, Sweden, Belgium, Who has the Best WEEE Implementation? (Paper presented at the International Electronics Recycling Congress on January 9-11 2002 in Davos)

Klemmer P (1994) Nachhaltige Entwicklung aus ökonomischer Sicht. Zeitschrift für angewandte Wirtschaftsforschung 7(1): 14-18

Klemmer P (1999) Nachhaltigkeitspolitik - eine ordnungspolitische Würdigung. In: Junkernheinrich M (ed) Ökonomisierung der Umweltpolitik. Beiträge zur volkswirtschaftlichen Umweltökonomie. Angewandte Umweltforschung 15. Berlin

Klemmer P, Löbbe K, Lehr U (1999) Umweltinnovationen. Innovative Wirkungen umweltpolitischer Instrumente, Vol 2. Berlin

Linscheidt B (1998) Ökonomische Anreizinstrumente in der Abfallpolitik Berlin.

Lucassen WHP (2002) Solutions on WEEE to the Industry. (Paper presented at the International Electronics Recycling Congress on January 9-11 2002 in Davos)

Metcalfe S (1995) The Economic Foundations of Technology Policy: Equilibrium and Evolutionary Perspectives. In: Stoneman P (ed) Handbook of the Economics of Innovation and Technical Change. London

Nelson RR, Winter SG (1982) An evolutionary theory of economic change, Cambridge, Mass., Belknap Press.

O.N. (2002) ZVEI: Kosten von 350 bis 500 Mio € pro Jahr durch Altgeräterichtlinien. In: EUWID-Europäischer Wirtschaftsdienst 12(51) S. 1

Pearce DW, Turner KR (1990) Economics of Natural Resources and the Environment. New York

Rogers EM(2003) Diffusion of Innovations, 5th edn. New York

Rogers EM, Shoemaker FF (1971) Communication of Innovations, A Cross-Cultural Approch, 2nd edn. New York

Skottheim J (2002) A Clear And Well Defined Producer Responsibility. Improving The Relationship Between Economy And Ecology. (Paper presented at the International Electronics Recycling Congress on January 9-11 2002 in Davos)

UN Commission on Sustainable Development (CSD) (2001) Indicators for Sustainable Development: Framework and Methodologies. New York

United Nations Division for Sustainable Development, Agenda 21. http://www.un.org/esa/sustdev/documents/agenda21/english/agenda21chapter4.htm

United Nations Division for Sustainable Development, Agenda 21. http://www.un.org/esa/sustdev/documents/agenda21/english/agenda21chapter21.htm

Zieschank R (2002) Umweltindikatoren im Kontext einer nachhaltigen Entwicklung. In: Zeitschrift für Umweltpolitik & Umweltrecht 4: 477-514

Water Management Towards Sustainability – An Indicator System to Assess Innovations[1]

Hartmut Clausen and Joachim Hafkesbrink

1 Introduction

Sustainable Development describes a long-term path for an economy or a part (e.g. a sector) of an economy that is consistent with particular – normative – criteria. While highly aggregated models often deal with just one indicator, namely consumption per capita, several indicators are necessary in order to describe a real path. The simple reason is that methodological problems and knowledge deficits impede among others a comprehensive monetary assessment of our doings at present and in future[2]. Therefore, the sustainability concept is frequently put in concrete terms by distinguishing economic, ecological and social aspects. Each of these three "pillars" in itself is often made up of a whole string of different aspects. Therefore the question arises what the idea of sustainability in context of water services means and by which indicators it can be expressed.

Beside the discussion of an exact definition of the term "sustainable water management", it has to be clarified how to reach this future requirement. In other words, those innovations should be realised that provide the highest contribution to the indicators chosen. Though policy makers need not necessarily know those, they should have a general understanding of the factors and (eco-political) instruments that induce firms to be innovative in a useful sense seen from an overall economic point of view. Ideally, it should be possible to convert those characteristics of innovation-friendly general conditions into indicators for sustainable innovation.

Thus, the aim of the paper is to make a proposal for a scheme describing innovations and their qualitative assessment with regard to sustainability (chapter 3). First we will outline the implications of the sustainability concept for the water services sector (chapter 2.1). Following that we will give

---

1 This work is part of the interdisciplinary research project AquaSus. We greatly appreciate the support from the Federal Ministry of Research and Education. The views expressed are solely those of the authors

2 The title of Jessinghaus' (2000) paper illustrates this problem, it reads: „On the Art of Aggregating Apples & Oranges“

a brief overview of typical indicators of innovation before we merge the two concepts of sustainability and innovation (chapter 2.2). After outlining the indicator concept in chapter 3, we explore some empirical evidence for the German waste water sector. In this context it seems useful to discuss briefly the regulatory and institutional environment that distinguishes the waste water sector from other sectors (chapter 4.1). Then some results concerning the determinants of innovations that stem from a survey conducted among German firms providing waste water services will be presented (chapter 4.2). We close with some political implications (chapter 5).

2 Sustainable water management, innovations and their indicators

2.1 Indicators for sustainable water management

Water resources fulfil several functions for both the ecological and economic system. Therefore the corresponding allocation problem has a quantitative and a qualitative dimension. Besides, present uses can be accompanied by irreversible damages to water resources. Thus, in a first step, it is obvious to put the idea of sustainable water management into concrete terms by using the so-called management rules for natural resources (Daly 1990; Enquete-Commission 1994). According to these rules, water uses should take place only at rates less than or equal to the natural rate at which they can regenerate. On the basis of this management rule, further principles were formulated (table 1) that have been – at least essentially – adopted by different organisations like The Federal Environmental Agency (Brackemann et al. 2001) and the professional association ATV-DVWK (ATV-DVWK 2001).

Table 1. Principles of sustainable water management

Principle of sustainability	Content
Regional principle	Regional orientation in the management of water resources; avoidance of interregional externalities (definition of regions in accordance to hydrological criteria).
Integration principle	Water is to be managed in its context to other environmental media. Integrated view of ecological, social and economic demands of the concept of sustainability.
Polluter-pays-principle	Allocation of cost of water use according to the use of water resources.
Co-operation and participation principle	Sustainability as common task of state and society. Public participation in decision making.
Minimisation of resource use principle	Reduction of resource use and increased use of regenerative natural resources.
Precautionary or prevention principle	Avoidance of measures with high potential of damage and/ or risk. In practice: principle of minimisation, put down in the drinking water ordinance.
Point of pollution principle	Prevention of a release of harmful substances at the place where they emerge, i.e. no end-of-pipe technologies wherever possible.
Reversibility principle	Consequences of measures in the field of water management should be reversible whereever possible.
Intergenerational principle	Taking the interest of future generations into account.

Source: in accordance to Kahlenborn and Kraemer (1999); Brackemann et al. (2001)

Some of these principles like the precautionary and prevention principle are already important for the German water policy[3]. The regional principle got higher importance by adding to The Federal Water Act (WHG, Wasserhaushaltsgesetz) the obligations to meet the needs of the public water supply above all by local resources and to control the uses of water re-

[3] According to INGU (1999), five of nine principles can be found in German water management laws. Of course, a mention in water laws does not ensure that these principles will be adhered to everywhere in an appropriate way

sources on the basis of river basins. Although these principles give – compared to the management rules – a bit more detailed view of the sustainability concept, some questions remain unanswered. Firstly, even if every single principle would be self-explaning, well defined ecological objectives remain still the prerequisite for following them. To this end the EU Water Framework Directive (WFD)[4], which came into force at the end of 2000, sets new yardsticks (Kaika and Page 2002). The WFD has introduced the "good status" of water bodies as a general aim of water management. Member states are required to achieve a good surface water and groundwater status by 2015. This means that surface water bodies shall have a "rich, balanced and sustainable ecosystem and that the established environmental quality standards for pollutants are respected". While physical-chemical criteria are familiar in water management, the ecological criteria are what can be called an institutional innovation. For groundwater bodies, a good chemical quality as well as a good quantity state (abstractions less than or equal to the rate of recharge) are required.

These objectives and the other requirements introduced indicate that the WFD tackles the main threats to sustainability with regard to water bodies: over-abstraction and pollution[5].

Secondly, the principles do not refer equally to all three "pillars" of the sustainability concept. Economic and social aspects are hardly explicitly mentioned. A very complex demand is entailed in the integration principle, according to which water management should be based on an integrated view of ecological, social and economic aspects of sustainability. Up to now, all actions like the reduction of pollution and measures of groundwater protection have been subordinated to the objective to supply drinking water of high quality[6]. Especially economic aspects like cost efficiency

---

4 "Directive 2000/60/EC establishing a framework for the Community action in the field of water policy", adopted on 23 October 2000

5 These changes concern, among others, the switching to a water management based on river basins, the so-called combined approach, the introduction of the costs of environmental externalities into water pricing and an increased public participation while developing water management schemes (Kaika and Page 2002)

6 See for example Ministerium für Umwelt und Verkehr Baden Württemberg (2000) and Niedersächsisches Umweltministerium (2002). The states Lower Saxony and Baden Wuerttemberg have called commissions of experts who formulated concepts of a sustainable water supply. However, in Lower Saxony the experts failed to reach consensus about the appropriate principles and proposals for the State water policy. Additionally, they were strongly focused on

have played a minor role in the water services sector so far. Only recently, there is a growing debate about appropriate measures for enhancing the economic performance of the firms (cf. WrcPLC and Ecologic 2002; Ewers et al. 2001; Brackemann et al. 2000). Although this has not resulted in substantial changes regarding the regulation of the sector and the degree of competition between firms, it might be seen as the beginning of a new defining process of the objectives of water management.

In order to clarify the relative importance of single objectives (e.g. a good drinking water quality, low prices and sufficient conservation), mechanism of participation of the public could be further developed. Public participation is a general demand in the context of sustainable development. The requirements made in the WFD (article 14) are rather a first step in this direction than a comprehensive scheme of participation of the public in water management decisions. The WFD only requires measures of information supply and consultation, whereas active involvement of the public is encouraged but left to the discretion of the Member States (ENGREF et al. 2003).

In any case and irrespective of any deficits, further criteria and indicators are needed which specify both the principles and the other aspects of a sustainable water management. Even though a whole string of sustainability indicators does exist (e.g. OECD 1998; BMU 2000), there is no uniform indicator system in use. Instead, scope and contents of indicators vary depending on both the actor who draws up a report and the addressee the report is for. Thus, as a first step the indicators and their desired level that shall represent the objectives of a sustainable water management have to be chosen. As a starting point we can partly fall back on standards defined in the different directives, acts and ordinances. Changes of these indicators in the desired direction indicate the impacts of innovations for which the drivers shall be identified and likewise expressed by indicators.

2.2 Indicators of innovation

Studies dealing with indicators for innovation focus foremost on so-called primary players like universities, research institutes or firms which invent new products, technologies or materials, take out a patent for their innovations or sell their new products. In this context, typical indicators are expenditures in R&D, education level of the staff, number of publications and patents, or the share of innovative sales at home and abroad. This ap-

the conservation of water as a regenerative resource and failed to take into account other aspects of sustainability plies essentially as well to studies on environmental-related industries, i.e. industries which produce environmental friendly technologies, etc. The study of Gehrke et al. (2002), for example, presents mainly the following innovation indicators referring to water management: Publications on environmental science, patents for technologies of water purification, production of technologies of waste water treatment, indicators of trade (revealed comparative advantage, relative world market share), turnover of products and services for water protection, public expenditures on environmental protection and environmental research.

Typically these indicators of innovation inform solely about inputs in innovation activities or about outputs of these activities. What remains often unexplained is to what degree certain inputs determine the outputs. Besides, most of the studies do not deal with the impact of changes of output indicators on sustainability.

However, how fast and to what extent objectives of a sustainable water management will be reached does not only depend on the innovative performance of universities, research institutes and technology suppliers but also on the behaviour of the users of technologies of water purification and treatment. These users are, apart from industrial firms, water utilities whose main objective is to provide the public with drinking water and to run sewerage networks and sewerage plants rather than developing and supplying new products. Regarding the three phases of innovation, they contribute to the diffusion of new technologies rather than inventing them. Water utilities in Germany typically have – in contrast to some big French water firms – no inhouse research department, even though it is quite possible that we could find co-operations with universities or research institutes aiming, for example, at the first application ("adaption") of a new technology. Therefore, other indicators are required namely such indicators that describe and determine the innovative performance of a firm or other players within the innovation system respectively.

Indicators concerning the innovative behaviour of water utilities might be, for example, the

- number of co-operations with suppliers of "green technologies", universities and so on,
- participation in work groups of professional associations,
- introduction of environmental management systems,
- the number of "new" water treatment technologies in use (indicating the speed of their diffusion) and
- (changes of) parameters describing the quality of cleaned water.

The indicators are just a few examples for indicators of innovation. A more systematic discussion of different types of indicators is subject of the following chapter.

3 Indicators for sustainable innovation

3.1 Indicator concept outline: multi-level analysis of the impacts of external drivers on innovation for sustainable development

This chapter defines a special indicator system to assess:

1. the impact of innovation drivers on the behaviour of actors (development and dissemination of innovations in the field of water supply and sewage management) (question 1 in figure 1),
2. the general importance of these drivers for the genesis and diffusion of innovations (question 2 in figure 1)
3. the possible impact of these innovations on sustainable development (question 3 in figure 1),
4. the status of sustainability indicators in the water services innovation system (WSIS) (question 4 in figure 1),
5. the factors that determine the dissemination of new technologies (complexity, compatibility, risk etc.) (question 5 in figure 1)
6. the general importance of these factors hindering the use of new technologies or organisational innovations (question 6 in figure 1).

The aspects number 3 and 5 are supplemented by additional evaluation tools to go into more detail concerning the leverage effects of each innovation issue and the diffusion criteria of innovations (see chapter 3.3 and 3.5).

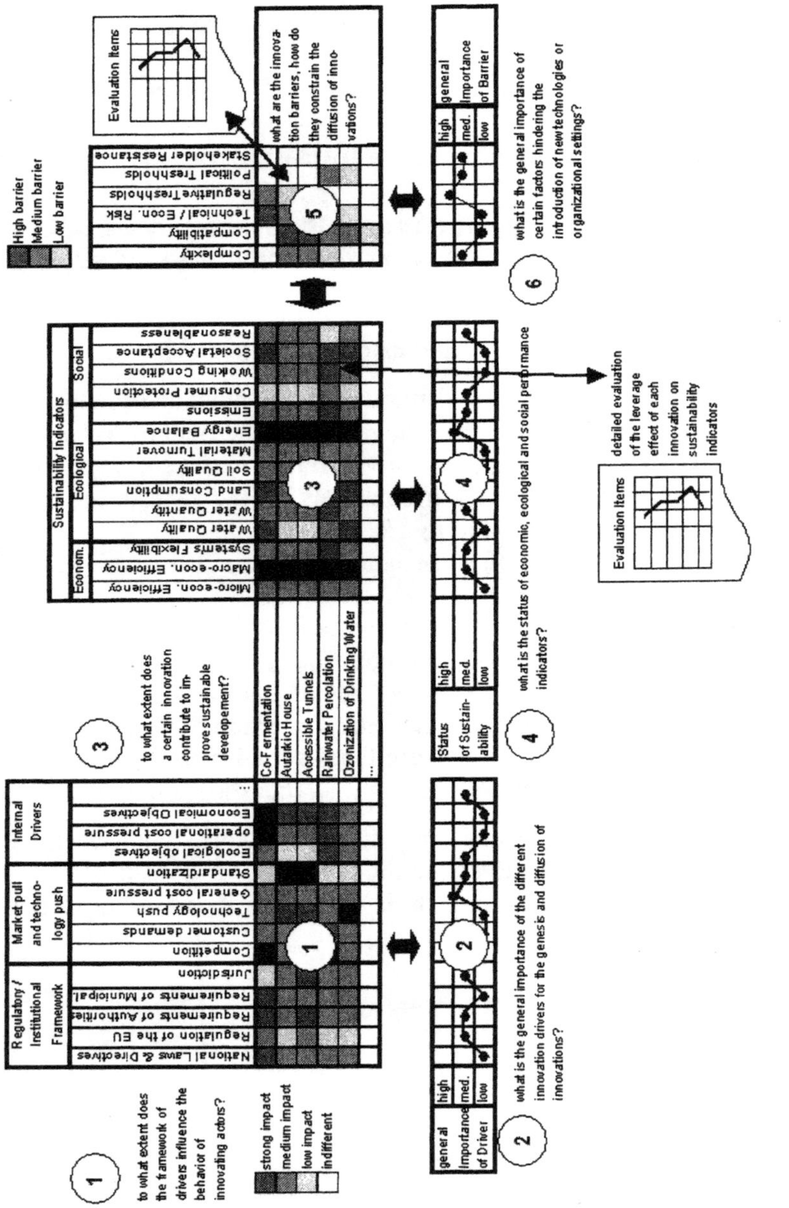

Fig. 1. Proposed Indicator System

The proposed system was developed primarily for the development of hypothesises for the AquaSus project. Secondly, the indicator system may serve as a corporate tool for decision making, since it is of utmost interest for water management companies to guess impacts on their incentive systems and to reach a clear understanding of further impacts of innovations on their business as well as on their economic, ecological and social performance indicators. Thirdly, the indicator system may be used by policy makers in environmental issues.

3.2 Impact of innovation drivers on the development and dissemination of innovations

Based on expert interviews different innovation drivers affecting the WSIS were identified. The elements of the regulatory framework, such as national laws and directives (i.e. the German Federal Water Act, Drinking Water Ordinance), municipal laws, waste laws on federal and municipal level, tax laws are supposed to have a strong influence on innovation behaviour. On the European level regulations such as the IPPC directive[7], the Water Framework Directive, laws governing material flows, bans of certain materials like phosphates, laws for plant protection, have to be taken into account. On the implementation level, requirements of authorities within the approvals of water management systems as well as requirements of the municipalities are also expected to influence the genesis and diffusion of innovations. Even the jurisdiction is supposed to influence the development of innovations, since it prescribes selective behavioural standards for innovation actors, i.e. in the adoption or implementation of technical measures.

Besides regulatory drivers, innovation is triggered by market factors and technology push phenomena. Especially on an international level, German water management companies are less competitive suffering from their fragmented structure compared to e.g. French companies like Suez Ondeo and Vivendi (Deutsche Bank Research 2000). Customer demands (demands for a certain water quality etc.) are expected to gain more importance in the future assuming that especially industry will ask for different water qualities for their processes. Pressures on municipal state budgets as the result of tax shortfalls etc. might force innovation on the supply side. Turning to technology, the availability of certain new technologies may also have an substantial impact on the implementation of new services and

7 IPPC Council Directive on Integrated Pollution Prevention and Control (96/61/EC) of September 24, 1996

management systems. Lastly, standardisation and institutional arrangements of norm-setting professional organisations affect innovation behaviour: a faster dissemination of new products and services can be expected when they comply with international standards (like ISO, PAS, EMAS, IPPC)[8].

Of course there are additional internal innovation drivers following the idea of a resource-based view and institutional perspectives of a firm as a result of intrinsic incentive systems such as intangible assets, internal sources, development needs of employees. Among these internal drivers ecological objectives (intrinsic ecological objectives as a result of image marketing etc.) may play an important role in the genesis of innovations, as well as pressure on running costs (demand to lower running costs after drinking water and sewage water prices have risen). Finally, economic objectives, e.g. being profitable, drive innovation to a substantial amount, at least in "normal" industries. These drivers may have an impact on certain innovations directly and selectively or indirectly in conjunction with other drivers in the innovation system. So there is empirical evidence of only those regulatory drivers having a serious steering effect on innovation behaviour (for instance the nitrogen legislation) which have their operational execution controlled by public authorities.

The drivers for innovation may be evaluated first according to their general importance on the innovation process by rating their general impact with "high", "medium" or "low". In chapter 4.2. we will give some empirical findings on these evaluations.

Moreover, within the indicator system, these drivers may be evaluated separately regarding their impacts on the development and diffusion of specific innovations (see chapter 3.3). The evaluation procedure uses the following categories of scores for each driver and each technology or innovation:

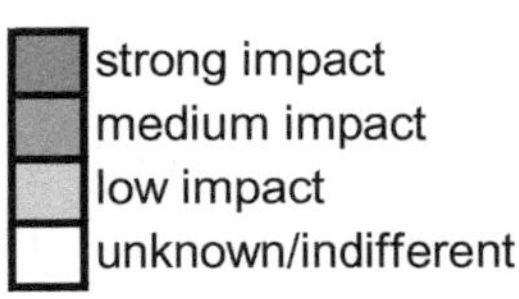

Fig. 2. Strength of impacts

At this point it should be noticed that innovation barriers are treated separately in the indicator system. Thus we differentiate impacts from

[8] ISO: International Standardization Organization; PAS: Publicly Available Specifications; EMAS: Eco-Management and Audit-Schemes; IPPC: Integrated Pollution Prevention and Control

drivers, whether they reveal with positive[9] or negative[10] incentives (related to question 1 of figure 1) or they reveal with innovation barriers, understood as a definite constraint for innovations (question 5 of figure 1).

These informations served as a basis for the items to be tackled in empirical investigations in AquaSus. The hypothesises were included in the questionnaires in form of special questions on innovation drivers (for results see chapter 4.2).

3.3 Impact of technologies on sustainable development

The next step within the AquaSus project was to develop a tool to assess the contribution of an innovation in the WSIS to sustainable development. Referring to question 3 of figure 1, the evaluation in this part of the indicator system follows again a simple process of estimating the leverage effects of each innovation or technology to the sustainability indicators. The scores are the same as in step 1 (see chapter 3.2).

In order to develop an easy-to-understand evaluation process, a tailor-made evaluation scale was developed for each sustainability indicator. These scales (see figure 3) may be called "qualitative scales", since they are non-metric scales, asking the innovation actors for a vote based on their understanding of the evaluation issue, for example specific norms. However, for the specification of these indicators especially in the WSIS and for the purpose of evaluating their sustainability effects, measurable indicators for water extraction and use, quality of resources and drinking water etc. exist. Nevertheless, it should be mentioned that for the specification of an indicator, there is no 1:1 relation between indicator and measurable ratio. On the contrary, different ratios and non-metric scales are often used depending on the kind and scope of the socio-economic empirical study. The indicators and their specific scales for the evaluation process are depicted in figure 3[11]:

9 For instance a threshold for particular substances determining the water quality

10 For instance a prohibition by law directing the innovation actor to search for other solutions

11 Several of these indicators are also part of the indicator system of the Commission on Sustainble Development (CSD), cf. BMU 2000

descriptor	scale			
economic criteria				
efficiency may be	not given	neutral	low	high
macroeconomical cost effectiveness	is reducing/ is getting reduced	remains unchanged		is improving
flexibility of system may be...	reduced	unchanged		elevated/ increased
ecological/ environmental criteria				
quality and clearness of groundwater	decreased/ reduced	potential insecure	neutral/unchanged	improved
quantitative condition of groundwater	decreased/ reduced	potential insecure	neutral/unchanged	improved
Land consumption	rises	unchanged		drops
quality of soil	drops	remains equal		rises
substance turnover	rises	remains unchanged		drops
energy balance	worsened/ deteriorated	remains unchanged		improves
air emission	rise	remains unchanged		drop
social criteria				
consumer and health protection	reduced	potential insecure	neutral	improved
work terms / conditions	worsend/ deteriorated	remains unchanged		improves
social acceptance	(organized) objection	low acceptance	high acceptance	(organized) consolation
reasonable for clients/residents/population	unreasonable (massive encroachment circumstances of life)	reasonable (change of the encroachment of life is social accepted)		no consideration/objection/, because there is no direct concernment
	Positive (desired) direction of the indicator →			

Fig. 3. Scales of sustainability indicators for AquaSus (comprising of two innovations for "Autarkic House" (broken line) and Co-Fermentation (direct line)

3.4 Status of sustainability indicators in the water services innovation system

The indicator system also helps to evaluate the status of sustainability indicators on a normative basis. Since the evaluation step in chapter 3.3 is concentrated on the leverage effect of single technologies or innovations on sustainability (question 3 of figure 1), the evaluation step according to question 4 in figure 1 is directed towards a general assessment of the sustainability indicators in the WSIS. This extension of the system is especially useful for investigations into the innovation system thinking of macro-economic modelling of sustainability indicators.

Even for corporate and political decision processes, the innovation actors may gain a more detailed view about sustainability when using a non-metric scale like "low-medium-high" as used in the AquaSus indicator system. These assessment may be based on normative borderlines like

- limiting values for hazardous substances in sewage discharges,
- average values for efficiency ratios both on macro- as on micro-economic levels, defining the average as "medium score",
- average values for land consumption, material turnover, air emissions and so on.

This – by the way – may be one of the major problems in evaluating the sustainability of a certain innovation, technology or even an entire innovation system: for all of these indicators absolute values may be developed based on primary or secondary statistics, measurements or other data sources. So the ratio "material turnover" might display a value of 200 m^3/inhabitant for instance for sewage sludge, but whether this should be judged as high, medium or low compared to a normative standard is not automatically answered. Within the empirical investigations only "relative evaluations" were undertaken to assess the contribution of certain innovations or technologies towards an improvement of the sustainability indicators. We followed the idea that – based on the principles of sustainability – an unmistakable "positive" and "desired direction" of the values of a specific indicator may be defined (see again figure 3).

However, looking at the indicator system in figure 1, for corporate and political decision makers, this might – as mentioned before – give hints for developing strategies on improving the sustainability of the WSIS as well. To keep it simple: just observe the potentials of the technologies rated at the heart of the indicator system (question 3 of figure 1) according to their supposed leverage effects on sustainability, and compare this with the choice for the status of a certain sustainability indicator. You may find those technologies or innovations which contribute very strongly on this single sustainability indicator by analysing the scores in the rows and columns of the matrix. Of course, this does not assure a "balance" in the development towards sustainability, because a single innovation or technology may contribute positively to ecological objectives but not to social or economic objectives.

3.5 Diffusion criteria of new technologies: innovation barriers analysed

Finally, the impact of an innovation depends on the scope and pace of its dissemination. With this background and looking on question 5 and 6 of figure 1, the next evaluation steps within the indicator system were to set up a tool to assess the diffusion criteria and innovation barriers for the innovations and technologies investigated in the AquaSus project.

The structure of these diffusion criteria rely on earlier works by Rogers and Schoemaker (1971) and Rogers (2003) defining the "diffusion" of an innovation as the process by which an innovation is communicated over time among members of a social system. According to these works the key indicators to explain the diffusion of an innovation are described as follows:

- relative advantage (degree to which an innovation is perceived as being better than the idea it supersedes)
- compatibility (degree to which an innovation is perceived as being consistent with existing values, past experiences and needs of the potential adopters)
- complexity (degree to which an innovation is perceived as difficult to understand and use)
- trialability (degree to which an innovation may be experimented with on a limited basis) and
- oberservability (degree to which the results of an innovation are visible to others).

In AquaSus a special adaptation method of these diffusion indicators was developed to cope with the tasks of the project. Following figure 1 the key questions to be answered were: What are the innovation barriers in the WSIS and how do they constrain the development and diffusion of innovations?

The question regarding the impact of selective diffusion criteria on single new technologies was specified with different indicators following the basic structure of the Rogers-Diffusion-Indicators, depicted in figure 4:

Descriptor	Scale			
type of innovation - productinnovation - processinnovation - organizational innovation (area)	innovation with changes/alteration in 1 area	innovation with changes/alteration in 2 areas		innovation with changes/alteration in 3 areas
technical innovation level	incremental improvement of one function	incremental improvement of several functions		huge improvement with changing the technology trajectory
risk and uncertainty of the implemenation for the innovating actor	low	medium		high
number of the involved protagonists in the implementation of the innovation	none	low, single level	middle, multi level, vertical or horizontal	high, multi level, vertical and horizontal
number of the persons concerned of the innovation	none	few participants	several	various/many
environmental media affected - groundwater - soil - air - waste	one media	two media	three media	four media
integration of substance streams in connection with utilization	not integrated (handling of substances is independent of utilization)	low integrated (handling of substances respects utilization method)	middle integrated (specific handling of substances with regard to possible utilization)	high integrated (definite orientation of handling of substances to a concrete utilization method)
segregation of substances	no segregation	low segregation	middle segregation	complete segregation
time slot of adoption	short-term (<3 years)	medium-term (3-8 years)	longer term (9-20 years)	long-term (>20 years)
time slot of diffusion	short-term (<3 years)	medium-term (3-8 years)	longer term (9-20 years)	long-term (>20 years)
necessary changes in the system of innovation system (as prerequisites for the diffusion of the innovation) see following indicators				
legal framework	not necessary	single issues	several issues	complex, multiple rule alterations
political support	not necessary	low	medium	high
necessary kowledge transfer to participants	not neccessary (knowlege exists)	low (only complement of existing knowledge)	medium (ambitious problem with new elements)	high (very ambitious and new problems in essential parts
creating a social acceptance for the innovation	not necessary	single group	several groups	all participants
(technical) interfaces need in the innovation system as a consequence of the innovation	not necessary	single technologies in interfaces	several technologies in addition to direct interfaces	complete transposition in all steps

Increasing complexity and risks as well as political/regulative treshholds, decreasing compatibility →

Fig. 4. Diffusion indicators for the Water Services Innovation System (Evaluation example)

According to the indicators displayed in figure 4, the diffusion of innovations in the WSIS is supposed to be more difficult,

- if the type of the innovation touches more than one area/element of the innovation system (i.e. the complexity rises if the innovation comprises both product novelties, process and organisational novelties)
- if the risk of the implementation increases for the innovation actor,
- if the number of innovation actors necessary to implement the innovation increases,
- if the number of environmental media affected increases (i.e. the complexity rises if the innovation touches at the same time groundwater, soil, air and waste questions and regulatory regimes).

These hypothesises were developed prior to the empirical investigations for AquaSus to structure the questionnaires in the areas of water supply and sewage water disposal.

The clusters displayed in figure 4 represent an analysis of diffusion indicators for the innovations

- Co-fermentation (of sewage sluge and biodegradable waste) (direct line)
- Autarkic House (self-sufficient, autonomous house concerning water supply and sewage treatment) (broken line)

based on expert interviews.

In terms of the general impact of diffusion barriers on the innovation process (question 6 of figure 1) some empirical findings will be presented in chapter 4.2.

4 Selected empirical results from the AquaSus project

The conventional notion of the term innovation as it is used for example in OECD (1997) refers especially to firms as innovating actors. Their management and performance depend according to the traditional industrial economics on basic conditions of the market's supply and demand side (e.g. technology, price elasticity, substitutes) and its structure (e.g. number of firms and customers, entry barriers, cost structures) (Scherer 1970). These issues are not only the outcome of market processes but result in many cases directly from the institutional environment that firms are embedded in. The national innovation system (NIS) approach takes the effects of institutions on innovations explicitly into consideration (Nelson 1993; Lundvall 1988, 1993). Furthermore, it is stressed that firms normally collaborate when innovating instead of undertaking all innovation activities within their hierarchy (Edquist 1997; OECD 2002).

With this in mind, we will briefly discuss the most important institutions, basic conditions and the market structure of the German waste water sector in the following chapter. After that the actual impact of single factors will be discussed (chapter 4.2).

4.1 The waste water services innovation system: market structure and regulatory framework

According to the German constitution (Art. 28 II Grundgesetz) waste water management is a sovereign task of the municipalities. They can decide between running on their own a waste water firm and delegating the task to independent service providers. However, they retain the ultimate responsibility which means they need to supervise the agent (Boscheck 2002). Only recently some few Federal States (Länder) allowed an assignment of the duty itself to private firms. Thus, Germany's market for waste water services is fragmented into many small "disposal areas" which are protected from competition by legal barriers to entry. These result from the fact that waste water services are classified by law as a sovereign task and that municipalities can force their residents to be connected to the public sewerage system and to enter into contract with the public firm ("Anschluss- und Benutzungszwang"). Thus, the municipalities play an important role within the innovation system (Clausen and Rothgang 2004).

While the markets are normally congruent with the administration districts, they can also consist of several municipalities that carry out waste water services in a common firm. Nevertheless, direct competition between firms, which is an important driving force for innovations, does not exist. Because of this and of certain rules for firms under public law, indicators like low prices, high profits or even the mere survival of a firm are not available to discuss the innovative behaviour of waste water firms. Anyway, those indicators would have only limited informative value unless external effects – the environmental benefit or damage of an innovation – are not fully internalised.

About 93 percent of the population are connected to the public sewerage network so that in Germany collecting waste water is a network-based industry (Statistisches Bundesamt 2001). The network generates about two third of the total costs of the disposal of sewage. This means that drain off waste water is not only characterised by legal but also by economic barriers of entry. These arise because pipe networks are accompanied by sunk costs, i.e. investment in pipes is neither recoverable nor can be used for other purposes (ENGREF et al. 2003). Sunk costs cause path dependencies and lock-in effects with regard to innovations (see Sartorius in this volume). As long as investments are not fully depreciated, a switch to another technology is more difficult than without sunk costs. How important this innovation barrier is in the individual case depends on the type (durability) and state of a certain part of a network and also on any facilities like sewage treatment plants that are complementary to a sewerage system.

The environmental regulation of the sector is dominated by command-and-control instruments even though taxes are levied on water abstraction and waste water discharges. In principle, all water uses require an official permission or a licence (§ 2 WHG). A permission to discharge waste water into rivers and lakes requires that harmful substances will be kept down as low as it is possible with water treatment techniques satisfying the "state of the art" (Stand der Technik). Before the WHG-amendment of 1996 only "generally recognised rules of technology" (allgemein anerkannte Regeln der Technik) were prescribed. The definition of uniform technological specifications takes place by close co-operation between representatives of water utilities, professional associations and the water authorities (BMU and UBA 2001).

However, against the background of the current regulations and the economic characteristics of the assets we can assume that the firms for the greater part of innovative activities stick to the well-tried technological trajectory. Though it is questionable whether a network based system remains flexible enough to deal with a declining water demand and the decrease in population in the long term. These developments might increase the pressure to innovate more radically. But the economic barriers to entry will not shrink until substitutes like small, decentralised cleaning systems are available with lower average costs than the current waste water fee. A prerequisite for the dissemination of those technologies is, on the other hand, that the legal entry barriers will be abolished.

4.2 Determinants of innovation

Finally we will present some results of a survey conducted among nearly 700 German waste water firms which are members of the "German Association for Water, Waste Water and Waste Services"[12]. We confined the analysis to three of the aspects introduced in chapter 3:

- the impact of factors belonging to the framework and market conditions on innovation,
- the role of the municipalities with regard to innovation,
- co-fermentation as an example of an innovation.

---

[12] Cf. Clausen et al. (2003) for a comprehensive description of the results

Asking the firms for their subjective perception of the importance of different innovation drivers, it turned out that the factor "cost pressure" is presently the most important driver from the firms point of view[13]. Almost 90 % of the firms which answered rate the total cost as 'very significant' or 'significant' for their innovation activities. Running costs are noticed to have a significant lower but nevertheless a high importance for innovation (position 4). Demands of authorities are judged as almost of the same importance as national acts and ordinances (position 2 and 3). Compared with these four factors, customer demands and competition play a minor role as innovation driver, they are placed at position 12 and 14. This is not surprising since direct competition, e.g. a choice between different suppliers, is missing. Thus, innovations are not particularly driven by the demand side of the market. Compared with these factors technology push is a bit more important: about every other firm judged the availability of new technologies as a very significant or significant driver (position 11). All in all especially costs aspects and the factors of the regulatory/institutional framework of the WSIS motivate firms to innovate. In future, regulation on the European level are supposed to have the biggest impact on innovation processes in the WSIS (58 of 261 answers) followed by "general cost pressure" (30 of 261), "running costs" as well as "national laws and ordinances" (both 25 of 261), "competition" (24 of 261) and "ecological objectives" (20 of 261). Among the future drivers "requirements of municipalities, "norm-setting authorities" are ranking low with 3 and 7 of 261 answers respectively.

Asking the firms to evaluate the importance of innovation barriers, it turned out that the cost aspect is regarded as the most important factor. The influence of some factors depends apparently on the kind of innovation. The same applies to the influence of the municipality on firms' innovation activities. On the one hand about 55 % of the firms have answered that demands of local politics do motivate their innovation activities (position 9). At the same time about 34 % of the firms take the view that the introduction of new technologies or organisational changes is hindered by local politics (position 2). Every fourth firm blames the local law for hampering innovations (position 3)[14]. A cost-cutting innovation might be promoted more likely by the municipality as an innovation that "solely" improves the quality of treated waste water.

---

[13] The factors were put in order by using the Wilcoxon Matched-Pairs Signed-Ranks Test

[14] Unfortunately the answer items of both question (motivation - obstacle) were not exactly the same and so the assessment of the overall effect of "local factors" is a bit difficult

The manner in which a municipality may hinder innovation activities can be illustrated by examples the firms mentioned in the survey. Several expressed that members of the municipal council were not willing to give up influence and are inclined to make decisions according to political reasons. Two firms stated that they could not introduce fee systems that realise the polluter pays principle better than the united fee for rainfall water and for waste water. Other firms criticised that their municipal council makes investment decisions conditional on initial set up costs while neglecting to look at long-term advantages and running costs.

Altogether, it seems possible to identify general determinants of innovation in the WSIS. However, direction and strength of the influence of a factor have to be judged for concrete innovations. Regarding co-fermentation it turned out that the expenditures necessary to introduce this treatment process for sewage sludge and the biological waste procedure are judged only by 17 % of the firms as "too high"[15]. Even though the technology does exist already for some years, nearly every third firm does not know about it. Besides, many firms believe that the technology has no advantages (23 %) compared to other methods and that it is accompanied by problems with the process technology (22 %).

5 Conclusion

The evaluation process within the AquaSus project shows, that – based on expert interviews prior to the empirical investigation – the concept of the indicator system can be used successfully to structure the discussion on the impacts of innovation drivers in the WSIS. Although not all of the elements of the indicator system were used in AquaSus, the system of interrelationships between innovation drivers, innovations, diffusion criteria and sustainability indicators shown in figure 1 contributed extensively to the process of formulating the hypothesises in AquaSus.

To conclude, a core set of substantial hypothesises can be derived from the experts′ votes for clustering the innovations according to figure 4 and the empirical investigation presented in chapter 4.2:

Thesis 1: The most complex innovations do not necessarily contribute to a greater extent to sustainable development in the WSIS than less complex innovations do.

Thesis 2: Even if the micro- and macro-economic efficiency of water management innovations is assured, the diffusion of the technology is not

---

[15] Nisipeanu and Thomzik (2004) explain the characteristics of co-fermentation and discuss any legal problems that may hamper the wide use of this process

necessarily assured, because the decision processes at the firm or municipal level might not be rational.

Thesis 3: According to thesis 2, a crucial innovation barrier might be the criteria of municipal decision making (investment costs instead average costs) as well as the conjunction with electoral cycles (increasing fees for water supply, sewage water disposal or waste management are unpopular).

From these findings we learned that our preliminary hypothesis has at least to be questioned: starting AquaSus with the general assumption that a quantum leap in sustainability may only be reached if comprehensive system innovations can be implemented, i.e. with the introduction of new integrated water service and sewerage systems that

- involve all actors in the innovation system,
- comprise product-, process and organisational change,
- lead to a new technological trajectory,
- integrate substance streams with a high orientation towards utilisation,
- reach a high segregation of substances,
- need alterations in the legal framework,
- has to create social acceptance in all parts of society.

The exploration of the in-depth clusters of technologies using the indicators of figure 1 (especially referring to questions 3 and 5) lead to a different result as stated in thesis 1-3.

However, coming back to methodological problems of the use of the indicator system, in our opinion the need of a comprehensive indicator system is obvious in order to assess contributions of innovations to sustainability. The examples investigated in AquaSus demonstrate that we will not be able to find a single indicator of sustainable innovations. As shown in chapter 1 and 2, the vision of sustainable development is rather complex. Since the environment of waste water firms in different regions may be different, not necessarily regarding the regulatory/institutional framework but with respect to the environmental/resource problems and the situation within the firm (organisational arrangement, culture), a comparison of patterns might help to reveal the main drivers of those innovations with predominantly positive contributions to indicators of sustainability. For these applications the indicator system may give a guidance for assessing systematically the chain from innovation drivers via incentives for innovations towards their contribution to sustainability. The indicator system presented in this paper may assist corporate decisions on these issues by asking a number of questions in form of indicators to be filled with data and information. In this context we learned that qualitative assessments of the impacts of innovations as made in chapter 3 are – in our view – not necessarily worse than quantitative assessments, even if it is desirable to assess the impacts in a quantitative way wherever possible.

As discussed in chapter 2.1. the details of an indicator system for sustainability are not worked out yet. Although principles of sustainable water management exist as indicated in table 1 and single indicators have been set up, so far we do not know the relationship between these indicators. The indicator system may help to agree on indicators of sustainability and on the values indicating the "Guard Rail", i.e. where to move without causing unreasonable environmental and sustainable impacts.

Hence, definitions and target values have to be defined on different levels (national, regional and local authorities). Mutual starting point for all actors in the WSIS is the legal framework. However that does not take everything into account. Waste water management firms – as a normative requirement – need to direct their innovation actions to those measures contributing to sustainability. Since the objectives given by the legal framework may be fragmented as the result of different legal regimes and overlapping ordinances, the targets so far may be too narrow and do not provide these 'Guard Rails' to sustainability. The indicator system presented in this paper may widen the scope of objectives by introducing driving, state and response indicators going beyond the medium water and asking for a transmedia-impact analysis of innovation measures in the WSIS.

Lastly, the indicator system may assist in setting up benchmarks between different companies or municipalities according to their innovation performance towards sustainability. Further research may be concentrated on the use of the indicator scheme in practise in order to reveal the patterns of successful innovations contributing to sustainability. These analyses should be conducted with the help of econometric methods based on survey data following the questions in the indicator system. The results of these investigations may acknowledge the so far qualitatively assumed leverage effects of water services innovations towards sustainability.

6 References

ATV-DVWK (Deutsche Vereinigung für Wasserwirtschaft, Abwasser und Abfall e.V.) (ed) (2001) Nachhaltige Wasserwirtschaft. Teil 1: Abwasser. Hennef

BMU (Federal Ministry for the Environment, Nature Conservation and Nuclear Safety) (ed) (2000) Erprobung der CSD-Nachhaltigkeitsindikatoren in Deutschland. Zwischenbericht der Bundesregierung. Berlin

BMU and UBA (Federal Ministry for the Environment, Nature Conservation and Nuclear Safety/Federal Environmental Agency) (eds) (2001) The German Water Sector – Policies and Experiences. Bonn, Berlin

Boscheck R (2002) European Water Infrastructures: Regulatory Flux Void of Reference? The Cases of Germany, France, and England and Wales. Intereconomics 37(3): 138-149

Brackemann H. Epperlein K. Grohmann A, Höring H, Kühleis C, Lell O, Rechenberg J, Weiß N (2000) Liberalisierung der deutschen Wasserversorgung. Auswirkungen auf den Gesundheits- und Umweltschutz. Text 2/00 des Umweltbundesamtes. Berlin

Brackemann H, Bartel H, Höring H (2001) Nachhaltige Wasserversorgung in Deutschland. Analyse und Vorschläge für eine zukunftsfähige Entwicklung. Berlin

Clausen H, Körkemeyer K, Lohaus J, Schroll M, Tauchmann H, Willms M (2003) Innovationen in der Abwasserentsorgung - Ergebnisse einer Umfrage. KA-Abwasser Abfall 50(12): 1563-1570

Clausen H, Rothgang M (2004) Innovations, Sustainability, and Efficiency in the German Water Sector: How to Push the Tanker into the Right Direction? In: v. Hirschhausen C, Beckers T, Mitusch K(eds) Trends in Infrastructure Regulation and Financing. International Experience and Case Studies from Germany. Cheltenham

Daly HE (1990) Towards some Operational Principles of Sustainable Development. Ecological Economics 2: 1-6

Deutsche Bank Research (2000) Wasserwirtschaft im Zeichen von Privatisierung und Liberalisierung. Aktuelle Themen 176

Edquist C (1997) Systems of Innovation Approaches - Their Emergence and Characteristics. In: Edquist C (ed) Systems of Innovation: Technologies, Institutions and Organisations. London

ENGREF Ecole Nationale du Génie Rural des eaux et des Forêts Montpellier (ed) (2003), Analysis of the European Unions Explicit and Implicit Policies and Approaches in the Larger Water Sector. (Work Package 1 of the EUROMARKET Project 'Water Liberalisation Scenarios'). http://mir.epfl.ch//euromarket

Enquete-Commission „Schutz des Menschen und der Umwelt“ des Deutschen Bundestages (ed) (1994) Die Industriegesellschaft gestalten – Perspektiven für einen nachhaltigen Umgang mit Stoff- und Materialströmen. Bonn

Ewers H-J, Botzenhart K, Jekel M, Salzwedel J, Kraemer RA unter Mitarbeit von Mankel B, Hansen W, Interwies E, Johannsen K, Scherer-Leydecker C, Voigtländer R, Herbke N. (2001) Optionen, Chancen und Rahmenbedingungen einer Marktöffnung für eine nachhaltige Wasserversorgung. Endbericht zum BMWi-Forschungsvorhaben 11/00. Berlin

Gehrke B, Krawczyk O, Legler H, Schmoch U. (2002) Umwelt und Wirtschaft. Dritter Bericht zur Umweltwirtschaft in Niedersachsen. Hannover

GG – Grundgesetz (1949) Grundgesetz der für die Bundesrepublik Deutschland vom 23. Mai 1949 (BGBl. I S. 1), zuletzt geändert durch das Gesetz zur Änderung des Grundgesetzes am 16. Juli 1998 (BGBl. I S. 1822)

INGU - Institut National du Génie Urbain Lyon (ed) (1999) R&D and Water Management Systems in a Perspective of Sustainable Development (Ger-

many, Spain, France, United Kingdom, Italy). Final Report for the European Commission, Directorate General XII, Lyon.
IPPC Directive (1996) Council Directive 96/61/EC of 24 September 1996 concerning integrated pollution prevention and control Official Journal L 257 , 10/10/1996
Jessinghaus, J (2000) On the Art of Aggregating Apples and Oranges. Nota di Lavoro 91.2000, Fondazione Eni Enrico Mattei. Milano.
Kahlenborn W, Kraemer RA (1999) Nachhaltige Wasserwirtschaft in Deutschland. Berlin
Kaika M, Page B (2002) Results of the Case Study on the European Union and the Water Framework Directive. In: Achieving Sustainable and Innovative Policies through Participatory Governance in a Multi-Level Context, Final Report of a Research Project funded by the European Community. Oxford
Lübbe E (2002) Jahreswasserbericht der Wasserwirtschaft. Gemeinsamer Bericht der mit der Wasserwirtschaft befassten Bundesministerien – Haushaltsjahr 2001. Wasser und Boden 54(7/8): 5-13
Lundvall BÅ (1988) Innovation as an Interactive Process: From User-Producer Interaction to the National System of Innovation. In: Dosi G, Freeman C, Nelson C (eds) Technical Change and Economic Theory. London
Lundvall BÅ (1993) User-Producer Relationships, National Systems of Innovation and Internationalization. In: Foray D and Freeman C (eds) Technology and the Wealth of Nations. The Dynamics of Constructed Advantage. London
Ministerium für Umwelt und Verkehr Baden-Württemberg (2000) Leitbild zukunftsfähige Trinkwasserversorgung Baden-Württemberg. Stuttgart
Nelson RR (ed) (1993) National Innovation Systems: A Comparative Analysis. New York
Niedersächsisches Umweltministerium (2002) Zukunftsfähige Wasserversorgung in Niedersachsen, Abschlussbericht der Regierungskommission. Hannover
Nisipeanu P, Thomzik M (2004) Das deutsche Umweltrecht als Einflussfaktor für Innovationen zum nachhaltigen Wirtschaften – Das Beispiel der Co-Vergärung von biogenen Abfällen in Faulräumen von Abwasserbehandlungsanlagen. Zeitschrift für Umweltrecht und Umweltpolitik 2/2004: 167-200.
OECD (1997) The Measurement of Scientific and Technological Activities. Proposed Guidelines for Collecting and Interpreting Technological Innovation Data - Oslo Manual. 2nd ed. Paris
OECD (1998) Water Consumption and Sustainable Water Resources Management. Paris
OECD (2002) Dynamising National Innovation Systems. Paris
Rogers EM (2003) Diffusion of Innovations, 5th ed. New York
Rogers EM, Shoemaker FF (1971) Communication of Innovations, A Cross-Cultural Approach, 2nd ed. New York
Sartorius C (this volume) Indicators for a Sustainable Technology Development – a Dynamic Perspective
Scherer FM (1970) Industrial Market Structure and Economic Performance. Chicago

Statistisches Bundesamt (2001) Statistisches Jahrbuch für die Bundesrepublik Deutschland. Stuttgart

WFD – Water Framework Directive (2000) Directive 2000/60/EC of the European Parliament and of the Council of 23 October 2000 establishing a framework for Community action in the field of water policy. Official Journal L 327, 22/12/2000.

WHG – Wasserhaushaltsgesetz (1957) Gesetz zur Ordnung des Wasserhaushalts. 27. Juli 1957. BGBl I 1957, 1110, 1386. geändert durch Art. 6 G v. 6. 1.2004 I 2

WRc PLC and Ecologic (2002) Study on the Application of the Competition Rules to the Water Sector in the European Community. Final Report

List of Contributors

Ahrens, Andreas
Ökopol - Institute for Environmental Strategies
Nernstweg 32-34, 22765 Hamburg
Germany

Beise, Marian, Dr.
Centre for European Economic Research (ZEW)
L 7,1 68161 Mannheim
Germany

Clausen, Hartmut, Dr.
German Federal Ministry of Education and Research (BMBF)
Hannoversche Straße 28-30, 10115 Berlin
Germany

von Gleich, Arnim, Prof. Dr.
University of Bremen
Badgasteiner Straße 1, 28359 Bremen
Germany

Hafkesbrink, Joachim, Dr.
ARÖW GmbH - Business and Consultants Network
Mülheimer Straße 43, 47058 Duisburg
Germany

Halstrick-Schwenk, Marianne
Rhine Westphalia Institute for Economic Research (RWI) Essen,
Hohenzollernstraße 1-3, 45128 Essen
Germany

Horbach, Jens, Prof. Dr.
University of Applied Sciences Anhalt
Strenzfelder Allee 28, 06406 Bernburg
Germany

Johnstone, Nick
National Policies Division, OECD Environment Directorate
2, rue André Pascal, 75016 Paris
France

Lutz, Christian, Dr.
Society for Economic Structural Research (GWS) Osnabrück,
Weissenburger Straße 4, 40076 Osnabrück
Germany

Meyer, Bernd, Prof. Dr.
Society for Economic Structural Research (GWS) Osnabrück,
Weissenburger Straße 4, 40076 Osnabrück and
University of Osnabrück
Rolandstraße 8, 49076 Osnabrück
Germany

Monßen, Melanie
University of Köln
Albertus-Magnus Platz, 50923 Köln
Germany

Rennings, Klaus, Dr.
Centre for European Economic Research (ZEW)
L 7,1, 68161 Mannheim
Germany

Sartorius, Christian, Dr.
Technical University Berlin
Einsteinufer 25, 10587 Berlin
Germany

Schleich, Joachim, Dr.
Adjunct Professor, Virginia Tech University, USA
Fraunhofer Institute Systems and Innovation Research (ISI)
Breslauer Straße 48, 76139 Karlsruhe
Germany

Walz, Rainer, PD Dr.
Fraunhofer Institute Systems and Innovation Research (ISI)
Breslauer Straße 48, 76139 Karlsruhe
Germany

FSC
www.fsc.org
MIX
Papier | Fördert gute Waldnutzung
FSC® C083411

Kolibri 360 GmbH
Ferdinand-Jühlke-Straße 7
99095 Erfurt, Deutschland
produktsicherheit@kolibri360.de